Chee

Cheese-making

Glynn Christian

Macdonald Guidelines

First published 1977
Macdonald Educational Ltd.
Holywell House, Worship Street,
London, EC2A 2EN

Made and printed by
Waterlow (Dunstable) Limited

ISBN 356 06018 7

Contents

Information

4 What is cheese?
7 Animals which make cheese
10 Why choose cheese?
12 The history of cheese
20 Produce and export
22 Who makes cheese?
24 Types of cheese
30 Large-scale industry
32 The traditional cheese-maker

Activities

34 Home cheese-making
46 Buying and storing
50 Serving cheese
53 Cheese and wine
56 Cooking with cheese

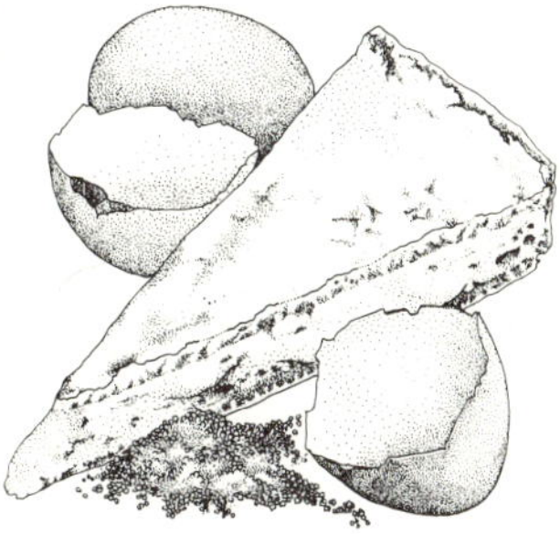

Reference

64 Guide to Europe's cheeses
66 French cheeses
74 British cheeses
80 Greek and Italian cheeses
83 German, Dutch and Swiss cheeses
86 Scandinavian cheeses
88 Travellers' cheeses
90 Suppliers and calorie chart
91 Books
92 Glossary
94 Index

What is cheese?

In German the word is Käse. In Dutch, Kaas. In Irish it is Cais, in Welsh Caws, in Spanish Queso, in Portuguese Quiejo. The Saxons knew it as Case or Cyse, and after the twelfth century it went on to become Cease, Coese, Schese, Chease, Ceise, Chies and Ches. Probably they were individual renderings of the same word all along. But in every case, they all meant the same thing: solid milk—or rather, solidified casein.

How dull it sounds! Yet legend and tradition maintain that cheese and cheese-making were the gifts of the Olympian gods to mankind. It was a royal repast in the Egypt of the Pharaohs; Zoroaster, the philosopher, lived for twenty desert-bound years on it.

The biblical references to it are legion—which leads us naturally on to the Romans, whose armies discovered new varieties all over their empire, and shipped them back to sell in the great Roman markets.

Cheese is as variable as the face of man and the places in which he lives. It can be a starter, a main course, a pudding, a savoury, a snack or an accompaniment to a drink. Like wine, it makes an ideal gift at a gathering of friends. There are probably over a thousand varieties of cheese made in the world today. In fact, this unnerving array breaks down into rather fewer basic groups, and the variations within these groups are often virtually negligible.

The basic principles by which milk is turned into cheese are common to all varieties.

◀ The veins in Roquefort cheese are thriving colonies of *penicillium roqueforti*. Once these invaded the "white" cheeses naturally, while they ripened in damp, dark caves. Nowadays commercial pressure means nothing can be left to chance, so virtually every blue cheese is inoculated with spores of mould. The same mould growth gives different-flavour results when grown in different basic cheeses—Stilton, Blue Cheshire and so on. Blue cheeses ripen from the centre outwards—as the veins spread.

Elementary facts

Milk straight from a cow or any other dairying animal will curdle unfailingly because of the action of the enzymes produced by bacteria naturally present in milk. These enzymes work on the lactose—milk sugar—converting it into lactic acid. This is what happens as milk ripens or sours. At a certain level of acidity the milk will coagulate, forming solid white curd and liquid whey.

This process of natural souring and coagulation can take some time and is affected by weather and other unpredictable factors. So it is often helped by the addition of rennet. This is made from the fourth stomach of a young ruminant—usually a calf—and its enzymes quickly curdle milk without requiring it first to be sour. This is an important way of influencing the eventual flavour of cheese—by starting with an acidic natural curd or a sweet, renneted curd.

Cheese-making was simple until the advent of pasteurization. Pasteurized milk will not curdle naturally. It may solidify if left somewhere warm, but this has nothing to do with the formation of lactic acid. Rather, it happens as a result of the

▲ Brie is a surface-ripened cheese, relying for the development of flavour on the growth of the white, furry *penicillium candidum*. Today it is sprayed on the young cheeses to ensure the correct results. Such cheeses ripen from the outside into the middle.

▼ This is a rare Farmhouse Scottish Cheddar. All such pressed cheeses rely on the action of enzymes to ripen them. They break down the protein and by-products of the action give flavour and aroma.

◀ Cheese goes through many changes before it appears at its appetizing best on our table. This rubbery-looking substance is curd during the process of cheddaring, an essential part of making many pressed cheeses. The slabs of curd are stacked on top of each other, and turned several times to facilitate even drainage of the whey and full development of the correct acidity. This simple technique was, surprisingly, perfected only in the last century.

action of potentially dangerous spoilage bacteria—solid or sour pasteurized milk can *never* be used for cheese. As well as killing off harmful bacteria, pasteurization also kills off the useful bacteria necessary for curdling milk and for giving cheese its aroma, flavour and texture. Nowadays a "starter"—a laboratory-grown culture of these cheese-making bacteria—must be put back into pasteurized milk if it is used.

Curds and whey

Once rennet-assisted or natural separation begins, the soft junket-like curds may be gently sliced and laid in a mold, while the whey slowly drains out over a period of days. Alternatively, the curd may be cut to release the whey more quickly, and then further cut or crumbled by hand. It may even be put through a small mill to break it down into very small particles. The amount of separation, the time it takes and the temperature at which it is done determine the final texture and flavour of the cheese.

Separation is usually accompanied by salting. Soft, naturally-draining curds will probably be sprinkled with salt as they lie shrinking in their molds. Drier curd will have to have salt mixed into it. These salted curds will then be transferred to a mold of some type and subjected to a greater or lesser degree of pressure. The amount of pressure and the length of time for which it is applied actively influence the type of cheese which results. Generally, soft, crumbly cheeses and blue cheeses do not have heavily pressed curds. The longer the curd is pressed, the longer the cheese will take to mature and the longer will be its life.

Ripening

After pressing, the cheese is put in a cool cellar or *cave* to ripen. Here, the catalogue of influences continues. The temperature, the amount of draught and the humidity of the ripening cellar all count, as does the treatment of the rind, if there is one, and how often the cheese is turned. This last process ensures even ageing through the entire cheese and, in particular, even distribution of remaining moisture. The progression to full flavour and market-readiness is supervised by some of the cheese industry's most important members.

Animals which make cheese

Cheese means milk, and to most people in Europe and America, milk means cows. If they are country folk, it may mean goats too. But other mammals, of very different appearance and habit, also provide milk.

Goats and sheep

Goats have been used as suppliers of cheese-milk from time immemorial. Their ability to clamber anywhere and live on almost nothing makes them a favourite possession of poor peasants and itinerant nomads. Because goats are normally tuberculosis-free and their milk-fat particles are smaller than those of cow's milk—making the milk easier to digest—they are also kept in large numbers in developed countries. In England, Her Majesty the Queen is patron of the British Goat Society.

Sheep, too, have long been traditionally associated with cheese. Nowadays French Roquefort, Italian Pecorinos and some authentic Greek varieties are the best-known cheeses based on sheep's milk. At one time sheep's cheese was among the most common types in England, made by monk and peasant alike. Many smallholders in France still make their home cheeses from sheep's milk.

Some unusual sources of milk

The Zebu—or Brahmin cow—sounds exotic and looks even more so. Its hump and drooping, fleshy neck are immediately recognizable, and although its home is India and Africa, it can be seen in Europe—and especially America—as part of breeding experiments. Crossed with either the yak or European/American breeds, it gives better milk yields and disease resistance.

The water buffalo can also be seen in Europe; in the southern provinces of Italy you may still find this "buffalo cheese" being produced.

Other, more unusual, sources of milk and cheese have become more the stuff of legend than reality. Laplanders still follow their reindeer over the steppes and smoke their hard cheese; but mare's milk—the original basis for Italian Caciocavallo and of a cheese dried on Scythian warriors' spears—tempts few palates today. Determined connoisseurs must conquer peaks and deserts to taste llama and camel cheese!

▶ Once regarded as sacred and virtually untouchable, these Indian Brahmin cattle are pictured at a Bombay dairy. Milk is becoming more widely accepted as a food in India and Brahmins are being cross-bred to give higher milk yields. Cheese, too, is now finding favour, much to the benefit of growing children.

The cows which are familiar to us are largely descended from the extinct Aurochs. From earliest times, man experimented with cross-breeding to produce cattle with specialist abilities. Some were bred to produce huge amounts of milk; others gave lesser amounts of richer milk. Today, most heavy milking breeds are unsuitable for beef and vice versa. But experiments are still in progress to produce a balanced animal which gives good yields of both.

Migration to the New World was particularly responsible for the development of new strains of cattle. The climate and prairie grasses of North and South America demanded hearty breeds. These evolved from the parent domestic cattle brought over with the first European settlers.

The variety of milk produced by these different breeds is just one reason for the world's astonishing array of cheeses. Each country usually favours a particular type of cow—one that produces a quality and quantity of milk suitable for the nation's cheese and dairy requirements.

In Britain, for instance, the herds consist of eleven different dual-purpose and dairy cows; but the vast majority—76 per cent—are black and white Friesians. By contrast, Friesians are relatively uncommon in New Zealand and Australia; the characteristic cow of this area is the soft-eyed Jersey. Its rich milk, high in carotene, gives Antipodean cheese a naturally golden glow and rich flavour.

Most European countries retain small numbers of traditional breeds, some of which are still valuable contributors. The British have a special affection for these old-timers; the British White, Welsh Black and the Kerry have tiny populations, while the lovely red Gloucesters, the original source of Double Gloucester cheese, now number only a few dozen.

Friesian The large-boned Friesian provides over 80 per cent of Britain's milk and is found all over Western Europe, the USA and Commonwealth countries. The milk production of Friesians is among the highest.

Guernsey Breton monks living on Guernsey in the Middle Ages bred the gentle Guernsey. She produces an average of 782 gallons (3,555 litres) per year of golden-yellow milk with creamy taste.

Normandy A hefty dairy breed, the Normandy is France's most common cow. It is thought to have been bred by the Vikings who, after invading France, crossed their own animals with the local breeds.

Ayrshire Originally bred in Scotland, this medium-sized, hardy cow gives milk exceeded only by Jerseys and Guernseys in richness. Ayrshires also comprise 10 per cent of the English herd.

Dairy Shorthorn This is a versatile cow that can be bred either to produce lots of milk or to produce top-quality beef, or both. It comprises 2-3 per cent of the British herd.

Jersey This big-eyed, small cow is always a favourite with children. Once called "Alderney", it produces almost as much milk as the bigger Guernsey, with a high butterfat content.

Braunvieh The famous Alpine cattle of the book *Heidi*. The Swiss Brown cow is a native of Switzerland, where it still predominates. It has also proved popular across the Atlantic.

Sheep In France the intelligent-looking Larzac sheep provides the milk for Roquefort; Italy favours the Sopravissina breed to supply the rich fatty milk for the famous Pecorino cheeses.

Goat Although French goat cheeses (*Chèvres*) are the best known, probably through being better distributed, many other countries have produced goat's-milk cheese since ancient times. With the present burgeoning interest in health, more and more people in the United States and Europe are taking an interest in the goat as a dairy animal.

Why choose cheese?

The affectation that regarded pressed cheese as an unacceptable ingredient in the diet of gentlemen—and a positive threat to the well-being of females and children—has long passed. Modern dairying, scientific analysis and plain good sense have restored cheese to its rightful place as a most important staple of our diet.

Cheese and your health

The blandest, immature, Cheddar-type cheese from the most unlikely source makes an important contribution to our daily intake of protein, calcium, Vitamin A and riboflavin, because cheese retains these valuable constituents of milk. The calorific value is usually high and more than 90 per cent of cheese is digestible material which may be used for growth and repair of body tissue or for energy.

Even with today's higher costs, cheese's penny-for-penny return in nutrition is quite astonishing when compared with other foods. Its versatility alone is unbeatable. It may be eaten raw or cooked, as part of a main meal or as a snack, at any time of the day. Older people usually prefer more mature flavours, while milder cheeses are usually suited to younger tastes. Babies can enjoy and easily assimilate grated, mild cheese from as early as nine months old. Some of the light and particularly digestible cheeses, such as Caerphilly, are irreplaceable components of invalids' diets.

The specific composition of cheese varies according to the milk from which it is made, but almost all is low in sugar—except for whey cheeses. As all the well-known English varieties are made with whole milk, they contain large amounts of the vitamins and minerals mentioned above and Vitamin D.

For dieters and slimmers

For those who wish simply to cut down on the amount of food they eat, cheese is the ideal starting point. Small amounts yield large proportions of our minimum daily intakes of many important health requirements. For years, cottage cheese has been regularly included in slimming diets. This cheese is basically the curd of skimmed milk that has been washed several times; the washing ensures that excess fat and the sugar-laden whey have been removed. But because cottage cheese is made of skim milk, it is not as high in food value as some other types. There are of course many other low-fat cheeses—Gouda and Edam are especially good in this respect. But remember that a food's fat content alone is not an accurate reflection of the number of calories it contains. The stated fat content of a cheese is shown as a percentage of dry matter, not of the cheese as a whole. So why does wet cream cheese have so high a fat content? Sometimes it is simply solidified double cream which has separated itself; in other cases the solid curd of whole milk has had cream worked back into it!

On page 90 is a simple table to give you an indication of how to calculate an approximate calorie count. There is no exact way, for as the moisture content of cheese varies enormously, even within the same piece of cheese, accurate assessment could only be possible if each mouthful were carefully analysed.

How the experts choose

In spite of modern technology, man himself still has an important contribution to make to cheese production. Specifically, this is usually associated with quality control.

Once a cheese has been made and put in the ageing room or cellar, expert eyes, noses, tongues, even knuckles, come in to play. Each cheese will be regularly checked, sometimes daily, to ensure correct rate of surface mould growth, rind formation and so on. Keen noses are needed to detect any unexpected spoilage, and special codes (bottom right) are employed to log a cheese's progress towards maturity. Cheeses with really tough rinds, such as bullet-proof Parmesan and Dutch Gouda, are tested by knocking on them with the

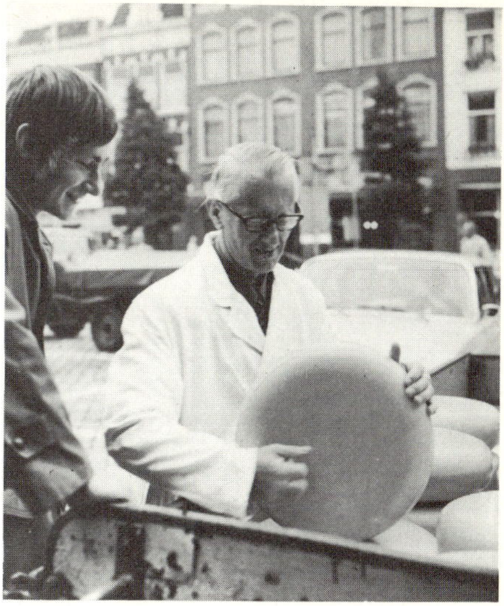

knuckles (above) or a special hammer. Years of training are needed before the sound can be deciphered correctly.

Tasting is simpler. When a long corer —called a trier or iron—is inserted and withdrawn, a sample of the whole cheese can be smelt and tasted (left). The core is then plugged back. Called "ironing a cheese", this test can still be requested in some stores when you wish to buy a whole cheese, such as a Baby Stilton or Cheddar for Christmas.

In many cases cheeses are bought young and then coaxed to perfection by a factor—the middleman of cheese distribution. Some factors simply buy and sell; others oversee bluing (Stilton or Cheshire), or ripening (Farmhouse Brie).

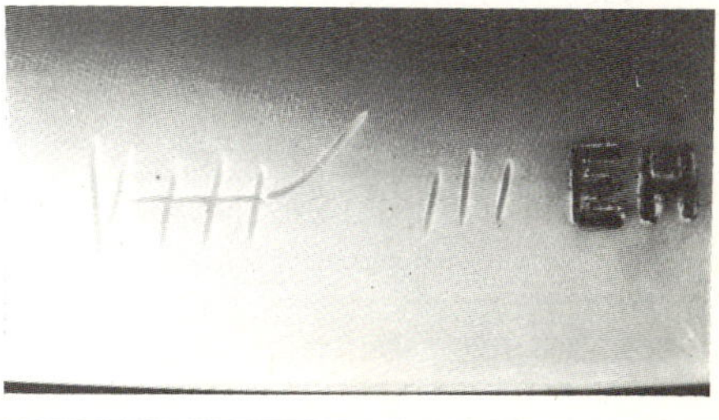

The history of cheese

Somewhere on the great, wind-ravaged steppes of Central Asia there should stand a monument as high as a mountain. For here, according to legend, once lived the simple nomad responsible for starting the whole business: cheese as we know it.

The earliest cattle to be domesticated were the Aurochs, a wild breed found throughout Europe and the Middle East. In Mesopotamia they often removed one of the animal's horns, probably for greater safety in the early days of milking. Cheese quickly followed cattle domestication, but only as the result of natural acid-curding, which took time and made an inherently sour-tasting product. One day, our unknown hero put his fresh milk into an improperly cured pouch he had made from the stomach of a young goat or sheep. Presto! the milk curdled in record time, and was still sweet.

Sweet, solid milk was a phenomenon the tribesmen had noted often in the stomachs of slaughtered, unweaned animals. Being necessarily thrifty they may have eaten and enjoyed it. Now they had the secret of making it themselves, and a wider world of cheese was created—a world that continues to grow today.

The spread of knowledge

That is only one version of the story. Some have it that the nomad rode with the stomach-bag attached to his saddle, others that it simply sat in the warmth of his tent. Still more reckon it was not like that at all; the Greeks swear that, like fire, it was a gift of their Olympian gods. Whatever its origin, no better substitute has ever been found for rennet, and rennet is still made from the stomachs of young suckling

▲ This fifteenth-century fresco from the Valle d'Aosta shows a variety of cheeses. Provolone types hang from a beam, rounds of pressed cheeses – possibly ancestors of Parmesan or of imported Swiss cheeses – sit on the counter.

◀ Cave paintings in North Africa have extended our knowledge of early dairy husbandry enormously. Dated about 5000-2000 BC, this one shows a full-uddered cow. Ancient peoples living in the area stretching from the Atlantic Sahara to pharaonic Egypt made simple cheeses.

mammals. Be it Brie, Chèvre, Roquefort, Gouda or even modern processed cheese, it all starts the same way.

Drifting nomads, conquering hordes, Roman legionaries, Christian monks and the written word slowly spread the knowledge until most of the world made cheese. There were some exceptions; it must be one of the few discoveries not credited to Chinese ingenuity. On the Indian continent, some religions taught that milk and cheese consumption robbed animals of their rights. In the Americas the Indians were hunters, not husbandmen; and on the Pacific Islands there were no suitable animals available.

Archaeologists continually astonish our

pasteurized society with more and more discoveries of ancient cheese-making knowledge. They cannot tell us how the cheeses were made or what they tasted like, but vessels depicted in sculpture and paint have been identified as ancient cheese molds and drainers.

Domestication of mammals for their milk probably started in Macedonia in about 6000 BC. Evidence of milk-processing a half-millennium later has been found in recently discovered cave paintings in the Libyan Sahara. This predates by two thousand years our knowledge of similar progress in both Mesopotamia and Pharaonic Egypt. The tomb of the Pharaoh Horus-aha yielded vessels containing a substance that scientists have identified as cheese.

By the time the Roman Empire stretched across Europe, cheese-making had progressed enormously and a wide variety of types was known. There were even alternatives to rennet, which used the juice of plants, flowers and bark to curdle and flavour milk. In most places goats were the primary source of milk; cows were not to replace them for some centuries, when the influence of Viking husbandry made itself felt.

The conquered and colonized had their own cheeses. Goths, Teutons and Celts made many varieties which the Romans thought good enough to import and sell in their markets. To the Romans, cheese was a staple. It was carried in every soldier's pack, sold in every market, mixed with honey, herbs, wine and spices for banquets; gratinéed dishes were an imperial

▼ Heating the milk, hanging curd to drain, hand pressing, pressing under stones— these eighteenth-century techniques are still used in farmhouses today.

For such an ancient, universal food, cheese makes only rare appearances in literature in comparison to, say, wine. Shakespeare mentions it only a few times, usually disparagingly. It seems cheese, especially toasted cheese, was notably inferior at the time and was eaten only in moments of great need. But the rotund Falstaff in *The Merry Wives of Windsor* looked forward to finishing his meal with "pippins and cheese".

Samuel Pepys's domestic staff often complained of the hard Suffolk cheeses they were given to eat, and were wont to name thin, miserable people "Banbury cheeses". But half-way through the seventeenth century, physicians were beginning to recommend certain types of cheese as useful in the diet.

speciality. The occupying forces passed on their own sophisticated cheese-making knowledge, too, and had a direct influence on much of the cheese we know today.

Vikings in Istanbul

After the Roman Empire went the way of the world, it was the unworldly brotherhood of Christian monasteries that cherished and preserved the cheese-making heritage of Europe for almost one thousand years. Simple, fresh cheese was an important part of a peasant diet, but large-scale manufacturing was attempted only by dairying monasteries. The Celtic monks of Ireland, unaffected by the waves of vandalism and ignorance that swept Europe after the Fall, were especially responsible for re-establishing the industry.

Most old-established cheeses in France, Italy and Switzerland trace their ancestry back to monastic beginnings, and to this day Port (du) Salut is made only by the monks of the Abbaye d'Entrammes.

The greatest single contribution to the renaissance of cheese-making came from a most unexpected source—the Vikings. It is not generally appreciated how wide was the sphere of their travelling, trading and influence. Some went as far as Istanbul, where blond bodyguards decorated the courts of Byzantine emperors. The Vikings used their trading routes to exchange expertise with the herdsmen of the Volga, the Black Sea, the Baltic countries, England and Normandy. They also shipped their cattle to new ports and evolved high-yielding cross-bred species that may still be seen today.

The clash of East and West during the Crusades from the eleventh to thirteenth centuries lifted Europe even further from her caseous Dark Age. The elegant and sophisticated dishes of the Arab and Byzantine world took Europeans by storm.

Today's Moroccan cuisine offers delicacies exactly the same as those claimed to be traditional Olde English, but which in fact had been brought back from the East, together with exotic fruits and flavourings. The new spices—cumin, caraway, pepper—found their way into both fresh and pressed cheeses. But it was the seductive combination of soft cheese with rose water or orange-blossom water that really awoke the European palate.

Pressed for trade

Big business and international marketing emerged with the establishment of commercial centres such as Venice and the cities of the Hanseatic League. Cheese was an internationally known and enjoyed food staple and thus an important part of trade. The boom in exporting led to a greater demand for cheeses which travelled and kept well—and hence to the creation of

CHEESE THROUGH THE AGES

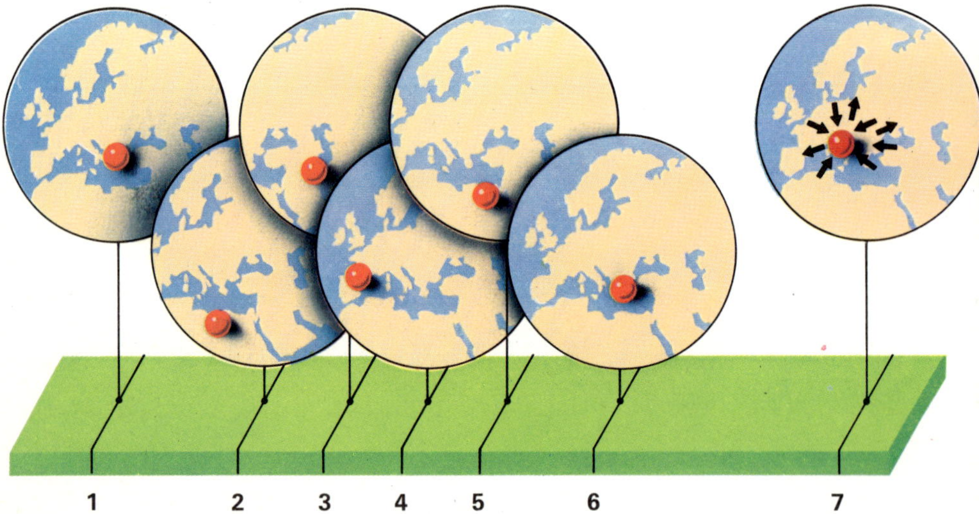

1 6500-5000 BC The earliest signs of dairying appear in Thessaly and Macedonia. It is not proven that cheese was also made.

2 5500-2000 BC Cave paintings in Chad and Libya show herds of cattle. They seem to have been used for dairy farming.

3 3500 BC Tomb art in Mesopotamia shows the milking of cows and storage of milk.

4 Pre-3000 BC The Basques—an isolated community, as today—already have words in their vocabulary for cheese and whey.

5 3000-2800 BC Egyptian wall paintings and pots of old cheese found in tombs provide first real evidence of cheese-making knowledge.

6 From 3000 BC The art of cheese-making spreads throughout Greece and Italy. Cheese written about and highly regarded by classical authors.

7 By 40 BC cheese from all parts of the Empire—particularly Germany—is brought to Rome. Cheese-making techniques are exchanged throughout Europe and Asia.

8 AD 500-800 Travelling Christian monks from Ireland encourage cheese production throughout Europe.

many pressed cheeses. The very fact that they were foreign was enough to recommend them.

Some cheeses were meant only for immediate consumption and could never travel; they required years of experiment and the advent of refrigeration before they passed into the common domain. It was only late in the nineteenth century that, for example, wooden boxes were invented for transporting Camembert.

You had to have your wits about you in the market place. Hygiene was known to have some bearing on results but there was not very much the maker or buyer could do about that. And there was no way of judging whether the memorable experience offered by mould or bacterial growth would be a taste thrill or an upset stomach. Temperature, acidity, curing and even pressing were arbitrary. Some say only one cheese in three was edible.

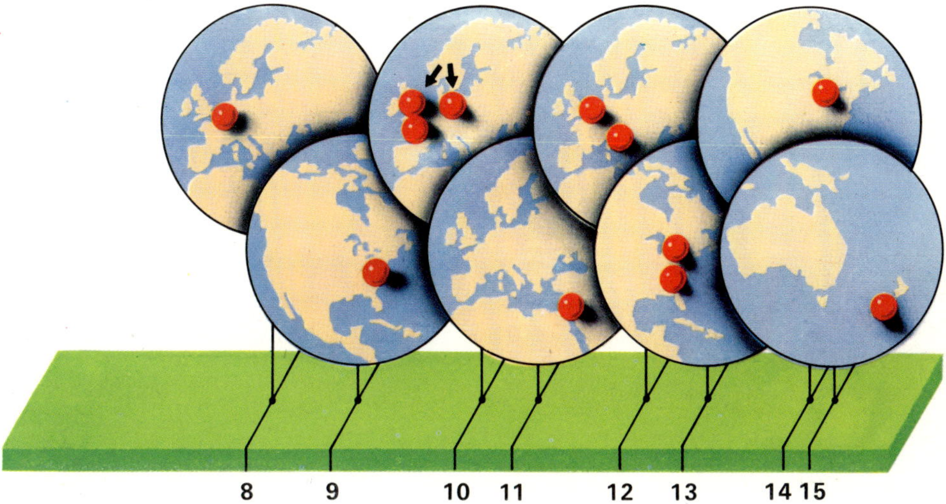

9 800 St Brendan takes cheese on his voyage to N America. Almost 1,000 years elapse before it is made there!

10 1000-1200 Voyages of Vikings to Britain, Normandy, Balkans and Black Sea; exchange of knowledge and breeding experiments.

11 1096-1270 The Crusades give Europeans a chance to sample cheeses from Greece, Turkey and the Holy Lands.

12 1400-1600 International trade centred on Venice and the Hanseatic League promotes higher standards and the standardization of many pressed cheeses.

13 1600-1700 Cows and cheese-making are brought to the new American and Canadian colonies by English, French and Dutch.

14 Post-1900 Scandinavian and German immigrants move to northern/Great Lakes area of USA. Particularly important in the development of a huge cheese-making industry there.

15 Post-1945 Descendants of European immigrants in Australia and New Zealand start up an ambitious cheese-making industry on a large scale.

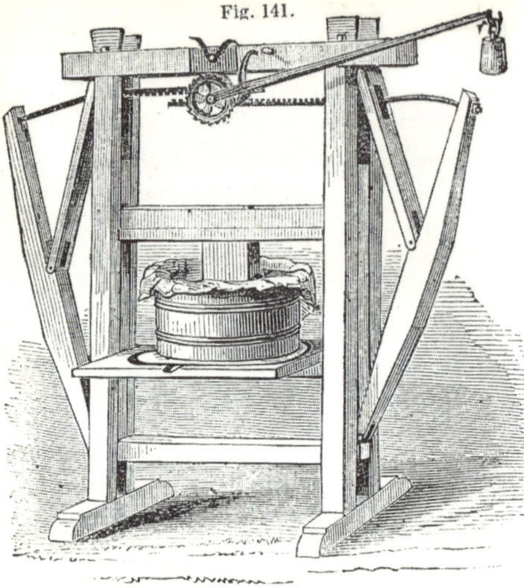

▲ A late-nineteenth-century cheese press offered for sale in America. The modern version is not much different.

▼ The drying room of an early cheese factory in Germany, not notably hygienic.

The beginning of the end

And so it remained right up to the early nineteenth century, when the new sciences first began to make nature a servant of man.

The work of von Liebig in Germany, Pasteur in France and Metchnikov in Russia explained the curing of cheese and established that by treating milk with heat you could render it pure and safe. It was a technique that spread as fast as the fevers and disease it prevented. Soon advances in sterilization made it possible to offer the world pure starter cultures to replace the dead bacteria in pasteurized milk. Rennet, too, could be standardized.

Burgeoning knowledge soon led to the harnessing of steam and the beginning of the end of manpower. By 1856 Norway was able to set up a cheese-making factory that was successful and profitable. Others followed all over Europe. In less than fifty years the cheese lore of thousands of years had been transformed, never to be the same.

The big cheese

Not surprisingly, the United States stands alone at the top of the big cheese league. Toronto's World's Fair displayed a four-ton monster, but that was well and truly capped by a Cheddar sent to the New York State Fair in 1937.

Among famous English cheeses was one given to Queen Victoria as a wedding present. Weighing about $\frac{1}{2}$ton, 9 ft 4 in (2·8 m) in diameter, and pressed between millstones, it was made from the full day's milking of the entire bovine population of East and West Pennard – some 800 cows.

After the young Queen had graciously accepted it, the cheese was sent straight back. It was the thought that counted; the present itself was not considered mature enough to be cut. Her Majesty promised £100 would be donated to the poor of the parish on the cheese reaching its majority.

Meanwhile, jealousy and commercial greed wrought their worst. A plaster-of-Paris facsimile was exhibited in London by a rival faction of farmers, and the original makers, not to be outdone, sent the Pennard Giant on a tour of the country. Lawsuits and injunctions followed. The Leviathan became a cheese in a gilded cage, to protect it from the ravages of hordes hired to hack it to death. Eventually, the battered prize came to rest in the sitting-room of an owner-farmer, an object of contention even among those who had a part in its being. It outlived its host, and crumbled slowly and ignominiously, food for pigs, not princes.

▲ A giant Cheddar made to celebrate the Silver Jubilee of Queen Elizabeth II. It weighs just over 100 lb. The tag informs us that it was made on 18 October 1976.

◄ The giant cheese presented to Queen Victoria for her Jubilee was imprinted by this handsome "follower", which fitted between the cheese and the press.

Overall cheese production (**bold** figure) and exports (*italics*) for 1973, in 1000s of metric tonnes. 1,000 tonnes= 904 UK tons.

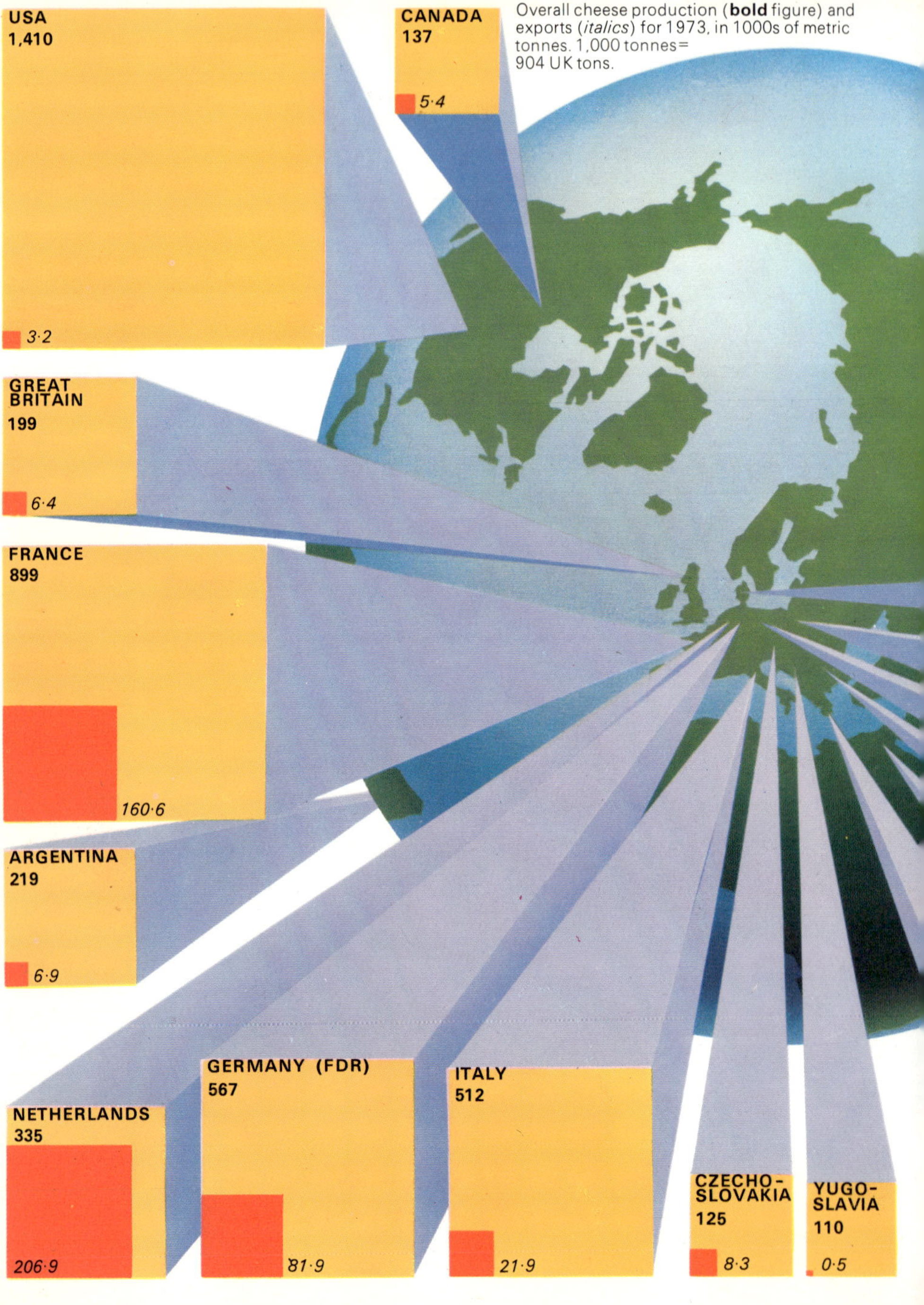

Produce and export

There is no global average proportion of cheese manufacture to cheese export. The extent of home consumption of national cheese product is controlled by the country's population, its eating habits and budgetary freedom. But exports are nearly always affected by the laws of importing countries.

From the strictly commercial point of view, many countries could never contemplate being serious exporters—success of a cheese type at home is never a guarantee of its acceptability abroad. While Cheddar and Brie have conquered nearly every market, curiosities such as Norwegian whey cheeses and German *Handkäses* have not.

It is interesting to compare the proportion of export to production in New Zealand and the United States. Both had no native dairying tradition but became major cheese producers relatively recently. Despite America's comparative nearness to the UK, New Zealand, as part of the old British Commonwealth, had protected entry to the English market and the United States was banned. Other ties are less formal but just as strong. Bulgaria and the USSR supply other Eastern countries.

Who makes cheese?

Apart from the Chinese, who have never regarded milk as quite suitable for human consumption and are nauseated at the idea of eating "rotten" milk, nearly every country makes cheese of some kind, even if it is not a long-standing tradition.

An international appeal

There are aberrants in China—the Mongolians make cheese from the milk from their prized mares, and the isolated region of Tibet is highly dependent on its supply of yak milk.

In India, greater religious freedom and dietary knowledge has made the production of cheese possible. Much of India's cheese is actually imported from Nepal, where a relatively new cheese industry has proved both successful and profitable.

Considering the fame of Canadian Cheddar and Wisconsin's huge output of cheese, it is astonishing to think that throughout the entire American continent milk was probably unknown as a human food until European settlement. But since 1600, North American cheese has pulled together the best of taste and technique from all over the world. Excellent pressed and soft cheeses trace their ancestry directly back to the ancient traditions of several European countries. The exciting mix of people and cultures, and the constant search for new techniques, has led to the creation of some remarkable American cheeses, such as Pineapple, whose hard rind can be used as a salad bowl when the softer centre has been scooped out. Processed cheese is also an American invention—one that has become an abused convenience world-wide.

In South America, fresh white cheeses are the most popular. Argentina exports a large amount of cheese and her Parmesan type has gained for itself a sound reputation abroad.

Australasia, particularly New Zealand's North Island, has perfect dairying climates. The most popular dairying cow there is the gentle Jersey whose carotene-rich, yellow milk gives the cheese excellent flavour, and high nutrition. Indeed, it has been reported that the Stilton made in New Zealand is at least as good as, sometimes better than, the real thing: quite a compliment, for there is barely a dairying country in the world that has not attempted the Stilton coup.

Perhaps the most interesting examples of modern dairy technology are Denmark and Ireland. Although both have been dairying countries for countless centuries, their current place in the world cheese market as manufacturers and exporters has almost nothing to do with their native cheeses. They provide excellent, if not quite authentic, imitations of the real thing. Some experts maintain that the Irish cheese industry faces the brightest future of any European producer, and that given economic and technical backing it could be turning out the best-quality cheese available in ten or fifteen years.

Danish Mozzarella, Irish Brie, French Cheddar and New Zealand Stilton have all been mastered. And what, you may wonder, do the clever Japanese do? Almost everything is the answer. A wide variety of the world's cheeses is made and marketed in Japan where, earlier this century, cheese was hardly eaten at all.

▶ The amount of cheese eaten in each country is an important reflection of the country's style and standard of living.

In France and Italy, high consumption reflects a peasant farming economy with high subsidies. The lower statistics of North America are justified by a wider, cheaper choice of protein products, while those of Finland are due to the scarcity of arable pasture and no tradition of cheese-making.

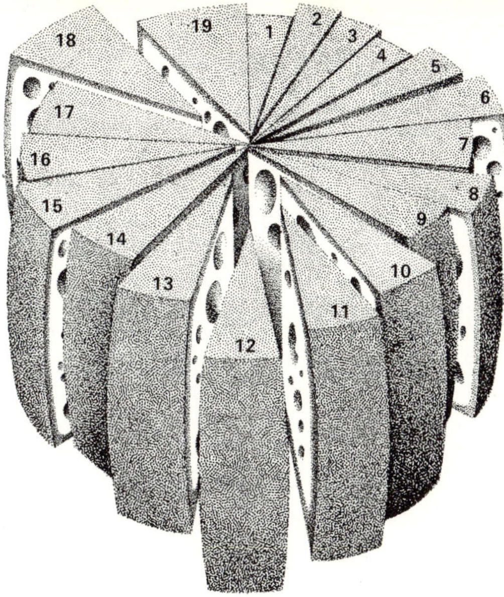

1	Great Britain	5·8
2	Finland	5·9
3	Australia	6·1
4	Austria	6·7
5	Canada	7·5
6	USA	8·6
7	Poland	9·1
8	Sweden	9·3
9	USSR	9·4
10	Norway	9·7
11	Belgium	9·8
12	Netherlands	9·8
13	Denmark	10
14	Switzerland	10·7
15	Germany (FDR)	11·1
16	Italy	11·2
17	Israel	12·5
18	France	14·8
19	Greece	15

Consumption in kg per capita in 1973. 1 kg = 2·2 lb.

Who eats it?

The French, surprisingly, do not top the cheese-eating scale—they are just overtaken by the Greeks. The French figures are quite understandable when you consider the wonderful range of delights they have to tempt them. But why do the Greeks eat so much cheese when they only have so little choice—no more than two or three types in most places? It is all a matter of national diet. Since meat is particularly scarce and expensive in Greece, cheese is the perfect protein substitute. Salty Fetta moistens the rather dry but filling Greek bread, is tossed in salad and decorates the local pasta in company with fresh lemon juice.

The figures on pages 20-21 are by no means comprehensive, but they do bring to light some very interesting facts. New Zealand, for instance, exports 90 per cent of the cheese she makes. But France's exports are higher than New Zealand's entire production, and the Netherlands exports a third more than France!

It would be most interesting to see a breakdown of national exports in terms of "native" versus "foreign" cheeses. For instance, does Denmark export more of her own Esrom, or more "Greek" Fetta?

If it were not for health regulations, which vary in each country, and complicated, strictly enforced quota systems, to protect native industries, these export figures might be even greater.

At present, however, a Californian wanting some good old-fashioned English Farmhouse Cheddar, or an Australian looking for a Reblochon Fermier, is probably doomed to frustration.

Types of cheese

▶ This small selection of cheeses shows examples of most of the main types. Type 1c is represented by peppered Boursin, 37, and garlic and herb cheese, 39. Type 2 cheeses are unpasteurized Camembert 17, Coulommiers on straw 18, Creamery Brie 19, Farmhouse Brie de Meaux 20, Neufchâtel 21. Type 3 are Bel Paese 15, Port Salut 16, Livarot 22, Petit Pont l'Evêque 23, Reblochon 24 (also lightly pressed). Type 4a is represented by Farmhouse Caerphilly 3. Type 4b types are Mature Gouda 1, Edam 2, Red Cheshire 4, Red Leicester 5, Sage Derby 34, Farmhouse Sage Lancashire 35, Double Gloucester with chives 36. Type 5 are Swiss Emmental 6 and Gruyère 7. Mozzarella, 14, is a plastic-curd cheese, Type 6. Gjetost, 33, is a typical brown whey cheese, Type 7. Type 9 cheeses include Roquefort 8, Blue Cheshire 9, Fourme d'Ambert 10, Blue Stilton 11, Bleu de Bresse 12, Gorgonzola 13. The big variety of goat's-milk cheeses shown are made in varying ways: 25 as Cendré; 26 soft, lightly pressed; 27 Mi-chèvre; 28 fresh, salted; 29 fresh, lightly salted, in red wax; 30 log with furry white mould; 31 chèvre pyramid with lighter surface growth; 32 Crottin—a harder, cured chèvre; 38 Poivre d'Auvergne is a Fondu or processed cheese, Type 11.

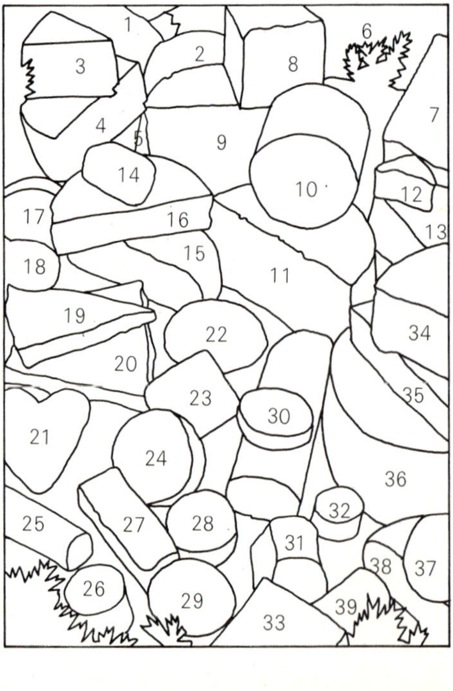

There seems to be no universally accepted method of classifying cheese into types. To do so by texture or taste is impossible; some cheeses change their textures as they mature and would belong to several classes. Taste can only be subjective. Appearance is equally deceptive—human whim, cunning and complication can produce cheeses which look like one thing but are quite another.

In this book, classification is by process of manufacture. Within each class, variation is infinite. Often a cheese's appearance gives some clue to how it was made. With guidance on this point, you should be able to guess at its condition and taste.

Fresh cheeses

These include cream cheese, curd cheese and cottage cheese. They may sometimes be lightly pressed after being salted and flavoured; Boursin is one such example. Although fresh cheeses are meant to be eaten fresh, this does not mean to say they may not have been matured for a short period. In former times this would be done between leaves or pewter plates.

This maturation is for development of flavour, not change in texture or growth of mould.

1a Cream cheese True double-cream cheeses need no help from rennet—the cream is allowed to solidify, then hung in very fine muslin and allowed to drain. More cream may be worked into the curd, to make double- and triple-cream cheeses. Erroneously, many cheeses made only of full-cream milk are defined as cream cheeses simply because of their creamy consistency.

Edam
ZNZ
de Hollande
NEUFCHATEL
FROMAGE MI-CHÈVRE
DEUX-SÈVRES
CHEF BOUTONNE
LE SOIGNON
FROMAGE DE CHÈVRE BLANC
PEU SALÉ
45
CEZAY
DENIZOT
CHAVIGNOL

1b Acid-curd cheeses "Acid-curd" indicates that the milk has been set by natural bacteriological action and the formation of lactic acid, rather than by rennet. Either raw milk, or pasteurized milk dosed with a starter, can sour and coagulate naturally. This natural time-consuming process is usually reserved nowadays for making soft, fresh cheeses (lactic cheeses).

1c Rennet-curd cheeses Commercial companies make most fresh cheeses—curd cheese, cottage cheese and so on—by renneting. Renneting ensures that the milk is ripe but not sour when coagulated. Cottage cheese is made from skimmed milk curds which have been washed and rinsed thoroughly.

Ripened cheeses

Ripening is often referred to as fermenting or curing. This is the largest and most diversified group; it is here that some cheeses escape definition by seeming to belong to several groups.

2 Soft paste cheese—bloomy, unwashed rind The curd for these cheeses is obtained at a relatively low temperature and never cut, other than when it is sliced and ladled into molds. There is no pressing: the whey drains naturally. Maturity usually takes about a month, and is accompanied by the growth of the *penicillium candidum* mould—a white fur. Brie and Camembert are the best-known varieties and the mould spore is usually introduced artificially by spraying. NB—these are generally creamy, mild, buttery cheeses. Once ammonia becomes even faintly evident they are overripe—the eccentric who eats Brie or Camembert when it is liquid or highly ammoniacal is foolishly endangering his health, and perpetuating the myth that this is how these cheeses should be eaten.

3 Soft paste cheese—washed rind The process is similar to that for Brie and Camembert, with the exception that the curd is broken up and stirred before being put into the mold. It is hardly ever pressed—Reblochon is an exception. Ripening takes from one to two months. During this time the rinds are frequently washed or rubbed and salted. This prevents the growth of mould but facilitates the growth of bacteria. The bacteria give these cheeses their easily recognizable rinds, which vary from a straw colour to brown-orange. Pont l'Evêque, Munster, Maroilles and Livarot are well-known examples of this group. They invariably have a rather high ammoniacal "nose", but, as often as not, are surprisingly sweet.

4 Scalded cheeses These are the popular semi-hard cheeses, a huge family made world-wide. These scalded cheeses are also called pressed, uncooked cheeses—a classification used as a contrast to the Emmental/Gruyère process. There are two main types of scalded cheese.

a Lightly pressed. The curd for all scalded cheeses is obtained at a much higher temperature than for soft cheeses. Separation of the curd and whey is achieved by several processes, including pressure and cheddaring (see page 33). The separation process is always done in a constantly maintained temperature, which might be higher than that for the curding—this is why the cheese is called scalded. When pressed relatively lightly, much whey is still contained, shortening the maturation period and limiting the life of these cheeses. Caerphilly and white Cheshire are prime examples of lightly pressed, scalded cheeses.

b Hard pressed. Exactly the same process as **4a**, but the scalded curds are subjected to greater pressure for extended periods, resulting in a firmer-textured cheese that takes longer to mature and has a much greater life.

The variations of process in making scalded cheese are innumerable, particularly with regard to the difference in temperature between that for curding and scalding. But this difference is never more than 10°C/15°F—if it were to be so the cheese would be cooked rather than scalded.

Scalded cheeses are given rinds in various ways, including brining (Gouda), spraying with hot water (commercial Cheddar) or wrapping with cheesecloth.

5 Cooked cheese After the curd has been obtained at still higher temperatures, the resultant mixture has its temperature raised even more—an increase of up to 20°C/40°F. This cooks the curd and produces a tougher, drier curd altogether. This is then pressed very heavily indeed. This process is used for Emmental and Gruyère-type cheeses. The dry curd and heavy pressing encourage longevity.

6 Plastic-curd cheeses The best-known cheeses of this type come from Italy—the smooth yellow Provolones, Caciocavallos and Caciottas. Their strange shapes hang like a schoolboy's store of tops and skittles and clubs in front of most Italian delicatessens.

The curd is soaked several times in hot whey which makes it very tough and putty-like. It is then kneaded by hand, making it even more malleable, until it can be molded and teased into the required shape. This type of cheese is also made in Turkey.

7 Whey cheese A prime example of "no waste, no want", these are not really cheeses at all, according to strict definition. Instead of being comprised mainly of solidified casein, whey cheeses are a solidification of the albumen of the whey. The solidification

◄ Lightly pressed and molded white cheese being "needled" with a culture of bacterial mould to induce the blue-green veins of Danish Blue.

Optimum usage of available feed is vital to a cheese-making economy. The lush alpine pastures of Switzerland (below) easily support high-yielding cows but would be wasted on low-milk-yield goats or sheep. The sheep of Roquefort (left) supply their fat-rich milk from comparatively sparse ground where cows would be next to useless.

happens when the whey is heated, causing the albumen to flocculate—collect in flakes or cloudy masses. In so doing, it picks up some of the remaining fat, vitamins and lactose of the whey. Ricotta and Greek Mitzithra are made like this. Nowadays milk is sometimes added to the hot whey to make a much richer end-product—similar to the next, much rarer, type of cheese, with which whey cheeses are sometimes confused.

8 Albumen cheese This is made by heating whole milk, which coagulates both the casein and the albumen. It is not very common but the Swiss Schabzeiger and Greek Galotiri are made this way. Both require some maturation.

9 Blue cheeses These are usually scalded cheeses that may have been lightly pressed. Light pressing, or no pressing, leaves minute spaces between the curd particles in which the moulds can grow. At one time the invasion of such moulds was by chance, relying on the effect of lack of hygiene, and the hope that most of the dairy and cheese-making equipment was infected with *penicillium roqueforti* or something similar. To hasten and ensure the process suitable cheeses are now inoculated with the bacteria by steel needles. Bluing should not be confused with ripening. A properly mature cheese might well be only slightly blue, because the mould did not grow properly. Conversely, an immature cheese may be very blue because it has been (purposely) injected with too much culture or because bad storage conditions encouraged its growth whilst inhibiting the cheese itself.

10 Goat and sheep cheese The complete range of cheese can be made with these goat and sheep milks, and many famous cheeses would once have been made with cow, goat *and* ewe milk as the season changed; both are specially suitable for home cheese-making.

Today goat cheeses are usually made like the "fresh cheeses" above, and have either the *penicillium candidum* white mould or ash covering. A few have washed rinds and are sold highly ripened, often also macerated. Although there is a farmhouse goat's-milk Cheddar-type cheese now on sale in England, this curd is not generally heavily pressed. Sheep's-milk cheese is usually kept soft for home consumption. In the production of genuine Greek Fetta, it is lightly pressed for ease of storage and selling. It is blued for Roquefort, and pressed and matured for the wonderful Italian Pecorinos.

11 Processed cheeses Once described as a triumph of laboratory technique over conscience, processed cheese is essentially an emulsion of second-rate cheese, flavoured, stabilized, extended, sterilized and wrapped so as to impede any natural, enzymic or bacteriological action that might conceivably give some identity or flavour. Nonetheless, it is enormously popular with children and unsophisticated palates—probably because of its blandness—and is thus an important source of protein for millions of people. Some Australasian and American varieties are specially suited to cooking and, in this guise, very welcome.

Fondu cheeses, which look like the French Tommes, and are supposed to, are sometimes rather good. The ubiquitous walnut cheese and the one coated with black grape pips are the most common.

▶ A treat to look for in Italy, Burrino is a plastic-curd type of cheese, hand-molded around a pat of fresh butter. This proves that even cheeses are not always what they seem!

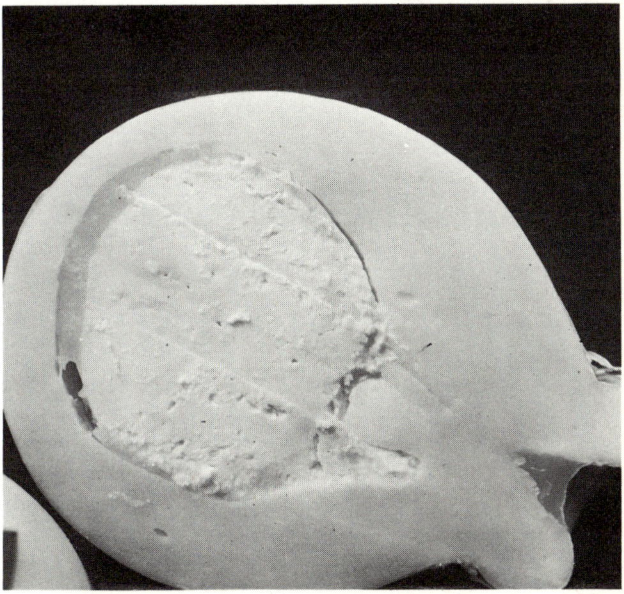

Large-scale industry

Pasteurization means the world is guaranteed vast amounts of cheese that is made hygienically and economically, but it also means that the richer individuality of the "little man's" product is missing. "Standardizing" ensures that the fat content of the cheese is consistent with its type requirements.

Commercial processes vary enormously. Illustrated here is a modern Swiss Emmental factory.

▶ Stringent tests ensure cheese quality; the Emmental stamp is a record of attention given.

▼ Most Emmentals are exported whole, but there is a growing demand for portions pre-packed in Switzerland itself.

▼ Now only time and a warm, humid ripening room are required for the Emmental to develop into full-flavoured maturity as the result of the action of bacterial enzymes.

◀ The curd is then cut into small pieces. This facilitates drainage of the whey. The mixture is further heated, or cooked, by slowly increasing the temperature to 127°F/53°C and stirring.

▼ Separating the curd from the liquid whey requires assistance. Cheesecloth on a flexible frame is tied, tightened over the curd and lifted, leaving the whey behind.

◀ Each Emmental cheese needs about 220 gallons (1,000 litres) of milk to yield close to 80 kilos of cheese. The raw milk is first heated to 91°F/33°C, then starter is added.

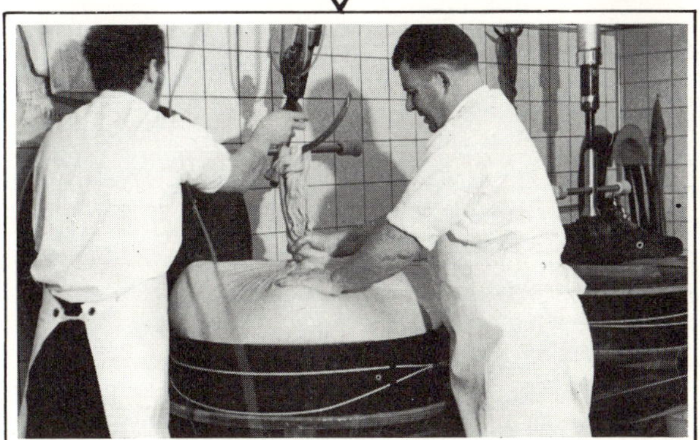

▲ The huge heavy bag of wet curd is allowed to drain further before being moved on to the mold and press.

▲ The fresh, compact wheels are next sprinkled with salt and put into a brine bath for several days. This helps form the rind which will seal in the cheese's flavour.

◀ At first the mass of curd appears too big for the press. But as pressure is increased it shrinks as more whey is expressed.

The traditional cheese-maker

▼ These nineteenth-century presses are still in use today in an English farmhouse dairy. Note the small truckle molds stacked on top of each other.

Traditionally-made cheeses fall into two categories—those produced for home consumption and those which are sold on a small scale. Commercial production of such cheese is widespread globally if you include the type of small manufacturing unit that neatly straddles both categories. The general store in a Greek mountain village will make quantities of Fetta and Mitzithra from the milk of the few sheep grazing on the hillside. Much of this cheese will be used by the proprietor; some will be sold to customers.

Small is best

Farmhouse cheese—whether Brie, Cheddar or Gouda—tastes different from farm to farm. An expert palate, primed with experience, is able to tell which region—even which farm—a cheese comes from. Because only the milk from a particular smallholding is used, it is flavoured with a mixture of herbs, flowers and water characteristic of its pastures. Cheese can vary even from day to day if the farm is big enough to allow cows to graze in different fields each day. And quality, of course, varies too. But though this statement is phrased in the present, traditional cheese-making is, in the main, a thing of the past. In the United Kingdom there are now only about fifty places still making traditional cheeses. Most is Cheddar and Cheshire, though the range is slowly increasing.

In France, even the most commercially exploited cheeses are still made on farms as well as in factories. These are the prized *Fermiers,* always more expensive, always looking a little more rugged, always using raw milk. In the pursuit and maintenance of things past, many farms guarantee to use only milk from pastures never treated with modern chemical fertilizers: truly, cheese "the way it was"—probably better if hygiene standards are taken into consideration.

Until very recently, even such a widely popular cheese as Roquefort was still based on farmhouse practice. The fresh white cheeses were made at home then taken to the communal caves at Combalou for bluing and maturing. Ever-increasing demand made this uneconomical, and now the Larzac sheep are milked mechanically in central depots.

Goat-cheese-making has resisted much of capitalism's inroads. Thus, even in France these cheeses are relatively expensive and in short supply. In the countries of the Middle East and North Africa, goat's cheese is a staple, usually made for early consumption from the family's own animal.

It might well take exceptional diligence, but you can probably find traditional cheese-making in most countries in Europe and Scandinavia. Searching would be specially easy in Holland, where a popular tourist diversion is watching flat, round Goudas being made in dairies attached to old wooden farmhouses.

Climate, lack of transportation facilities and a dearth of refrigeration guarantee the survival of traditional home- and farm-based cheese-making in hotter and less developed countries. Sadly it is in the so-called developed countries of north-western Europe that these crafts are being allowed to vanish.

► Even limited farmhouse production of Cheddar uses huge quantities of milk. Top: the curds and whey pouring into draining tanks. Middle: cheddaring and milling the curd to ensure proper drainage and acidity levels. Bottom: the pressed cheeses are sealed with lard and bandaged.

Home cheese-making

Any exhortations to return to the soil and make your own cheese, to get back to "self-sufficiency" and the real thing, must be tempered with a full understanding of two points: it is not as simple as it sounds, and it is nearly always more expensive than buying cheese, unless you actually own milk-producing cows. Everything from the time of the year to the temperature of your room and the type of cow the milk comes from will affect your cheese-making. But once you've made successful cheese, perfecting your own recipes and techniques, you'll be hooked.

There is no such thing as a definitive cheese recipe. The amateur cheese-maker will get different results from day to day. Experience will teach you how to adapt and adjust, so that even if your method varies the result will be (almost) consistent. The best way to start is with the soft cheeses. They give a greater yield per quantity of milk, are relatively simple and do not require expensive equipment. Although hygiene is of the greatest importance, the majority of soft cheeses, being eaten relatively fresh, are not as prone as some types to contamination and spoilage. Soft cheeses that require longer than a week to ripen need a high standard of hygiene.

For your first attempts, choose cheeses that are made with small quantities of milk.

◀ Difficulty of access once made home cheese-making imperative in the Orkney Islands. Today, creameries produce cheeses like this for distribution all over Britain.

INGREDIENTS

Milk Autumn milk gives the heaviest yields, even when pasteurized. But as far as flavour goes, pasteurized milk varies little all year round, and in any case you have to add a starter or buttermilk to restore flavour-giving bacteria.

If you use raw cow's milk you must be absolutely sure that the herd and cow are accredited tuberculosis- and brucellosis-free; also that none of the donating cows have been given any antibiotics during the previous three or four days, as these will kill the bacteria in the milk. Sheep's and goat's milk does not normally need to be pasteurized, although current thinking on the subject is best checked out with such authoritative bodies as the British Goat Society.

Milk with a high fat content, such as Jersey and Guernsey, does make richer cheese, but is more difficult to work with. It requires more rennet and higher temperatures. Stick to the basic silver-topped milk at first. It is most important to understand that pasteurization kills all the harmless lactic acid-producing bacteria. Thus, when such milk turns sour and curdles, it is because it has been contaminated. *Pasteurized milk which has "turned" must never be used for cheese-making, no matter how sweet it smells.*

Starter This is where cheese-making can become daunting. "Starter" is sterile milk injected with a controlled culture of bacteria to replace those destroyed by pasteurization. Though various bacteria govern cheese attributes ranging from flavour to holes to aroma, those which are of most concern to the home cheese-maker are the bacteria which form lactic acid.

Commercial starters are quite expensive and not readily available to amateur enthusiasts. But if you wish to do it the professional way, write to one of the suppliers listed at the back of the book, mentioning the type and amount of cheese you wish to make. When you make a purchase, the supplier will provide full instructions. He will probably tailor both the starter and his advice to your needs.

But there is an easier way to start: use cultured buttermilk. This is widely available, cheap and gives excellent results. Buttermilk is produced by adding a culture to the by-product of butter-making. The bacteria are almost identical to those used in cheese starters.

Rennet You must use proper cheese rennet, which is stronger than junket rennet, for most cheeses. With some experience, you can increase the proportion of junket rennet and successfully use this, but it is unreliable. Keep rennet dark and refrigerated—it will lose only up to 10 per cent of its strength a year.

Rennet has two main functions: to curdle fresh, unripened milk without souring it, thus providing a sweet curd; and, in cheese made from ripened or sour milk, to hasten the solidification process without further altering the acidity. The use of smaller or larger quantities of rennet results in softer or moister curd, which affects the speed at which the whey drains, which in turn affects the texture and flavour of the cheese.

Annatto This is cheese colouring and nothing else will do. Ordinary food colourings do amazing things—a beautiful orange stripe in a delicious York cheese turns a hideous purple after maturing has taken place. Be warned!

Salt This brings out the flavour of your cheese and also acts as a preservative. Ordinary household salt can be used: iodized salt is not recommended.

Lard Good pure lard is needed to coat pressed cheeses. This seals them while they are curing.

Confidence A good cook has been defined as someone who knows how to hide mistakes. In cheese-making you cannot hide many, but you may be able to ignore most: for as long as your hygiene is reasonable, your product will be edible and you will know better next time. Your first cheeses may not be what you intended. The secret is never to tell the others.

EQUIPMENT

The better your equipment, the better your results. On the other hand it is sometimes nice to be able to blame your tools, in the event of a real disaster!

Thermometers Ideally, use a floating dairy thermometer. Otherwise a cooking thermometer will do, but you will have to hold it, which can be tiring. A room thermometer is important, too, for the ambient temperature does have a marked effect on results.

Measuring jugs These should be of glass or stainless steel: plastic can retain odours, and if it is scratched it will be difficult to sterilize.

Mixing bowls The capacity of these depends on your ambitions and your pocket. Even small soft cheeses need a lot of milk: 1-2 gallons (4·5—9 litres) is a typical starting point. A bucket or two might therefore be in order. In any case, never use a metal other than stainless steel. Unchipped enamel is good and there is a cheaper, lighter special dairy bucket of plastic.

Large stainless steel bowls are a good investment. You will be surprised how useful they are if you do a lot of cooking and preserving.

Waterbath Milk is more successfully heated indirectly. The temperature at which you rennet most milk is low enough to be obtained by standing your bowl or bucket in hot water in a sink or bath and stirring. For higher temperatures milk heated slowly in a bowl or bucket directly on your cooker—but protected by a couple of asbestos mats—is perfectly suitable.

Spoons, ladles and knives Apart from teaspoons and tablespoons for measuring rennet and annatto, you will need something with a long handle to stir the milk, a perforated spoon or slice, a ladle for transferring curd and a long-bladed knife for cutting curd. Stainless steel is ideal.

Colander A large colander, even of plastic, is essential.

Molds Irritatingly, most traditional molds are no longer available, other than Coulommiers hoops, which can be used for several cheeses. You will have to improvise with tins of varying shapes and sizes, details of which are given with the relevant recipes. Watch for quick deterioration though, and remember always to punch any holes from the inside or you will never get your cheese out again. Some people have successfully used plastic piping cut into suitable lengths: you may think of other ideas.

Drainage boards Many cheeses drain on wooden boards and a basic small wooden chopping board will do as long as it is not pitted or scored. Make sure it is big enough to hold the number of molds you will use at any one time. In many cases—and always when using the Coulommiers hoops—you will need two.

If you are lucky enough to be able to get hold of some stout, stiff straw mats from your local cheese shop, you may be able to make do with these balanced over your whey collector.

Draining mats Table mats of wooden strips or straw, cannibalized bamboo curtains, or plastic-strip table mats can all be used. Cheesecloth, folded a couple of times, can be used *in extremis.*

Whey collector The drainage of cheeses results in a lot of free whey. You will need to stand your mold on its drainage board and/or mat in something such as a large roasting pan. See page 45 for making whey cheese.

Cheesecloths Use cheesecloth or butter muslin, both easily available. You nearly always use it double.

A cheese-making book You can buy one from most suppliers. It is used for recording *every* step in making every different type of cheese—from the age of the milk to the room temperature, the time that elapses between each operation, the acidity and so on. You will then have a record of your mistakes, your corrections and your triumphs. It may seem pedantic but it pays huge dividends.

Home-made bread and butter, home-grown herbs and vegetables—perfect complements to your own soft or pressed cheeses. The typically simple press is shown with metal and plastic molds and a two-tiered Coulommiers mold. The wrapped cheeses are truckle Cheddars, in front of which are a Coulommiers type, slices of Caerphilly and pressed goat's cheese, and herb-flavoured soft cheese.

TEMPERATURE

The room temperature for cheese-making is all-important. The best range is 65°-70°F (18°-21°C). Otherwise, decrease the milk renneting temperature by the number of degrees by which your room exceeds 65°F/18°C.

STIRRING

"Deep stir" means stir right to the bottom of the container in wide circles.

"Top stir" means agitate only the top $\frac{1}{2}$ in (1 cm) of the milk. This stops the cream rising, thus protecting flavour and texture.

RECOGNIZING COAGULATION

The easiest way to recognize coagulation is to float a piece of straw or a toothpick on top of the milk. Stop stirring the moment it stops moving or you will break up the curd and release whey far too early.

CURD-CUTTING

To test the state of curd, insert an index finger into the coagulum and move it forward an inch or two. The curd should split cleanly and should not adhere to your finger. The whey released should be clear and greenish. Cloudy whey means you are losing fat.

The time-honoured formula for checking when curd should be ready to cut is: *time taken to coagulation + 2$\frac{1}{2}$ times that time.*

If your cheese milk took 15 minutes to coagulate, it should be ready to cut in 15 minutes + 37$\frac{1}{2}$ minutes, i.e. 52$\frac{1}{2}$ minutes after adding the rennet. If ready to cut earlier, your curd is higher in acidity than expected; if later, it is lower.

Take great care to observe instructions as to the size of the curds. If cutting soft curd, wait a few minutes between each change of direction in cutting: this will help retain the butterfat.

Sterilization and washing

Everything you use, including working surfaces, must be sterile, which may be achieved by boiling for at least 20 minutes or using a sterilizing solution—1 fluid oz (28 ml) ordinary household bleach (hypochlorite) to 1 gallon (9 litres) water. Rinse very thoroughly.

All equipment that has been in contact with milk must be rinsed with cold water before washing with hot, otherwise an undesirable film will form.

Yourself Your hair, if long, should be tied back. Never wear anything woolly, for fine, hairy strands will *always* find their way into the cheese. Remove rings and ensure nails are clean.

TESTS AND TECHNIQUES

Milk acidity

Normally, milk for the pressed and matured cheeses should have an acidity of ·21-·23%. Soft cheese uses unripened milk, whose acidity need only be ·18-·20%.

Test 1 Measure 56 drops fresh rennet into a warm cup, add a straw or a toothpick and then 4 fluid oz (114 ml) milk at 84°F/28°C, noting the time exactly on a watch with a second hand.

Stir for a few seconds, then note the exact time the straw stops moving.

The shorter the time to coagulation, the higher the milk's acidity, e.g.:

24	seconds	·20%
21-22	seconds	·21%
20	seconds	·22%
18	seconds	·24%

NB If milk varies more than one degree from 84°F, it will affect the result by 2 seconds for every degree. Milk at 83°F will take 26 seconds to indicate ·20%, milk at 85°F will take only 22 seconds, and so on.

This test does not convert easily into centigrade, but for use as a very rough guide, the milk should be at 28°C and fluctuations of ·5°C will affect your result by 2 seconds; i.e. milk at 27·5°C will take 26 rather than 24 seconds.

Test 2 This is much more accurate without being much more difficult. Add 2 drops phenolphthalein to 10 cc of the milk or whey to be tested. Fill a well-calibrated plunger (acidometer) with N9 caustic soda solution; both chemical liquids are available at most chemists. Slowly drip the caustic soda solution into the milk or whey until it turns and *remains* pink. Each cc of caustic soda used shows the presence of ·1% acidity. Thus 2·5 cc=·25% and so on.

If you think you have too high or too low an acid count, you can compensate.

If development of acidity has been slower than required, for soft unscalded cheeses you simply have to let the milk sit longer; for

▲ If you start by making soft cheeses, which you should, you will find you have most of the equipment needed, or can easily improvise.

At the back: a large plastic bucket, or stainless steel bowl, in front of which are an ordinary plastic colander and glass measuring jug, and a roasting tray to catch whey as it is drained.

To the left and right of this group are some molds. You will have to buy the Coulommiers mold on the left; the other two are improvised, one from a large can with holes punched *from the inside out* (very important), the other a $\frac{1}{2}$-lb tea tin with the entire top and bottom removed. Baking tins, springforms, plastic tubes and ordinary cups and mugs can also be used, depending on the type of cheese and amount of drainage required.

The front group shows cheesecloth, a room thermometer and a floating thermometer, a wooden board, long sharp knife, scoop and measuring spoons. The straw mat needed to assist in drainage also allows of improvisation with table mats and so on, but folded cheesecloth can always be used.

pressed cheeses, heat the curd more slowly and to a lower temperature than usual to give slower drainage.

For soft cheese, fast-working curd must be cut small or sliced thin to hasten drainage; for pressed cheese, it is scalded at a higher temperature and taken up to that temperature as fast as possible.

POSSIBLE PROBLEMS

Mould can be wiped off with a damp cloth; dry cheese can be mixed with cream or butter; and wet cheese should be eaten fast, before it sours.

Failure to coagulate

Often caused by residue of sterilizing solution, wrong or weak rennet; in unpasteurized milk it may mean antibiotics are present.

Tainted flavour Invariably this is from unsatisfactory kitchen or personal hygiene.

Cheese which will not ripen You have probably killed the bacteria in the curd by contact with sterilizing solution (usually in the cheesecloth, which should have been boiled).

SOFT CHEESES

By the time you have worked through this range of soft cheeses you may consider yourself experienced. It goes from simple curd flavoured with citrus juice or vinegar to traditional molded cheeses. There is plenty of opportunity for individual expression, too.

Lactic curd cheese

An unrenneted, creamy cheese with endless possibilities. A gallon of milk makes 2-2½ lb (·8-1 kg) curd, but you can halve or quarter the quantities.

1 gallon (4·5 litres) milk
1 fl oz (28 ml) buttermilk
salt

Slowly heat milk to about 76°F/24°C, then add cultured buttermilk. Deep-stir well, cover the milk, leave for up to 24 hours at a room temperature of 65°-70°F/18°-21°C.

When the curd is firm enough to pass the "clean finger" test, line a bucket or bowl with a large, dry cheesecloth, pour in the curd, pull up the corners of the cloth and tie them tightly. For best results, allow 36-48 hours' drainage in a temperature of 65°-70°F/18°-21°C, and mix the dry outer curd with the soft inner curd every 3-4 hours.

Salt to taste—about ½ oz (14 g) is usual—then decide what to do with your cheese by looking at the ideas on page 41. Beating will make it creamier in texture.

Acid-curd cheese

3 pints (1·7 litres) full-cream milk (or milk and cream)
1 fl oz (28·5 ml) lemon, orange or lime juice *or*
2 tablespoons flavoured wine-vinegar, e.g. chilli, garlic, tarragon or cider vinegar

Heat milk just to boiling point, then stir in citrus juice or flavoured wine vinegar. Curdling is almost instant. Chop lightly, drain in cheesecloth. Orange- or lemon-flavoured curd makes a good cheesecake. To every 8 oz (226 g) curd, beat in 1 oz (28 g) butter, with sugar, currants, peel and nutmeg to taste. Bake in a pastry case. That's real home cooking!

Rennet-curd cheese
–basic milk cheese

2 pints (1·2 litres) pasteurized milk
½ fl oz (14 ml) buttermilk (optional)
junket rennet

Heat pasteurized milk with buttermilk. Slowly raise the temperature to 100°F/38°C, stirring all the time. Now make a junket with junket rennet (NB *not* cheese rennet). Stir well, cover, and leave in a warm place for 2 hours or until the finger test is successful. Line a colander with your cheesecloth and ladle thin slices of the junket into the muslin. Lift the four corners and tie the bag tightly with string. Let the whey drip into a bowl until drainage has stopped.

Remove it from the cloth and wrap it in waxed paper —cereal-packet liner is perfect.

It is ready to eat now, but is probably better left for a day or two. Sprinkle with salt and eat it with fruit, cream and sugar; or make it savoury (see right). If you want to be true blue and British, cure the cheese in nettle leaves or between pewter plates.

Single-cream cheese

1 pint (·6 litre) single cream
¼ fl oz (7 ml) buttermilk
14 drops rennet

Gently heat the single cream to 75°F/24°C and add the buttermilk. Mix well. Cover the mixture and leave to ripen for 2-3 hours. Then add the rennet, diluted with six times its volume of cold boiled water; stir it in very well. Cover and leave at a temperature of 70°-80°F/21°-27°C for 8-12 hours or until the cream coagulates. Ladle this curd into folded cheesecloth and hang it to drain in a cool, airy room.

At 4- to 8-hourly intervals, scrape the harder curd from the cloth and mix it with the softer central curd.

When the curd is thick and grainy, add salt. Mix well and shape by hand or pack into a yoghurt container or teacup. Serve sweet or savoury (see below).

Double-cream cheese

This uses the same proportions of cream and buttermilk as the recipe for single-cream cheese but is neither ripened nor renneted.

Heat the cream to 75°F/24°C, add the buttermilk and at least ¼ oz (7 g) salt, stirring in well. Cover and leave in a warm place to coagulate.

Mold and flavour it exactly

Flavour variations

Soft cheeses can be infinitely varied. Here are some ideas. Others will no doubt occur to you.

1 Work whipped cream into your drained curd, perhaps flavouring it first with garlic, vanilla, spices or rose petals.
2 Shape a small cheese, then roll it in sweet paprika, mixed fresh herbs with garlic and lemon or crushed green peppercorns. Always leave to cure a few days so that the flavour permeates.
3 Wrap an unsalted cheese in thin slices of smoked bacon.
4 Soak raisins in black rum then press the swollen raisins into the cheese.

The finger test (far left): moving your finger forward should result in a clean split, with no clinging particles.

◀ Cutting the curd should be done smoothly; rough handling can release too much whey.

▶ Curd and whey separation is often started by ladling the mixture into cheesecloth over a colander or bucket.

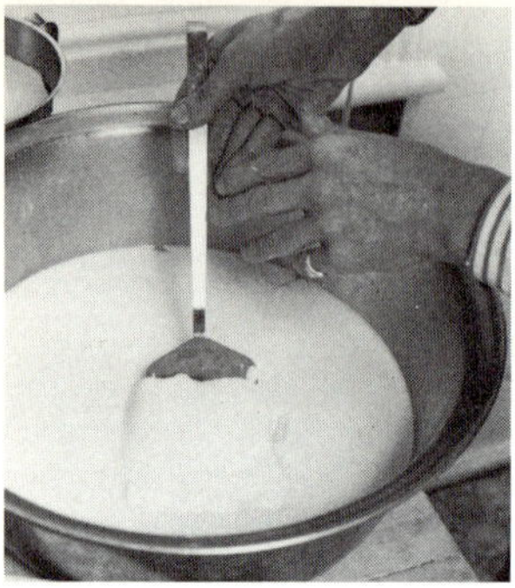

▲ **Making a top**: by pressing your mold into the curd, you can remove a perfect slice to reserve for the top of your cheese.

▲ **Filling the mold**: an ordinary "slice" makes fast work of filling a cheese mold, here standing on a straw mat to assist drainage.

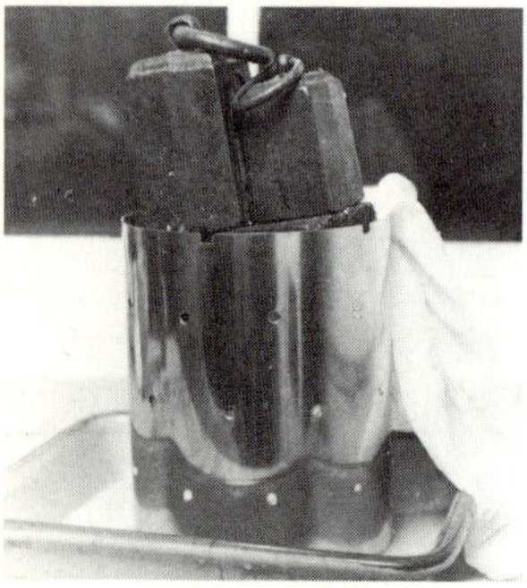

▲ An improvised cheese press.

the same as single-cream cheese, but with even more care, in view of its cost.

Coulommiers-mold cheese

This recipe will yield three cheeses (Coulommiers molds cost about £1 each).

1 gallon (4·5 litres) milk
2 fl oz (57 ml) buttermilk
56 drops cheese rennet

Heat the milk (cow's to 90°F/32°C; goat or sheep milk only to 85°F/29°C). Remove from heat, add cultured buttermilk and stir well.

Dilute cheese rennet with six times its volume boiled cold water. Deep-stir for a couple of minutes, then top-stir until coagulation. It takes at least an hour before the "clean finger" test works.

Always prepare "tops" for molded soft cheese, by gently pressing the mold into the curd, scooping out the shape and slipping it on to a clean plate; make one top for every cheese.

Stand the molds on mats on drainage boards and ladle slices into each mold in turn, so each is filled to the same height.

If your milk was too acid, cut thinner slices to hasten drainage; thicker ones impede it. Let the curds settle, then put on the three tops.

Drainage in a warm place will take about 24 hours. Then remove the top part of the hoop and turn the cheese in the bottom of the hoop on to a clean mat and board.

Repeat this process every day until the curd is firm enough to leave the side of the mold—usually three days. Then rub about $\frac{1}{2}$ oz (14 g) salt into the surface of each cheese.

Wipe off any mould or leave it on just one to see what flavour change results.

Now you have three cheeses to taste. Eat one fresh, leave another for a couple more days, the third a bit longer, wrapped in waxed paper in the refrigerator. Then you will know which flavour you prefer.

When you are adept, try putting layers of fresh herbs or pepper between the layers of curd.

Colwick cheese

A traditional English cheese that does not need turning. It is therefore easier to use an improvised mold. A gallon of milk makes two cheeses.

The recipe is exactly the same as for Coulommiers-mold cheese. But this time the mold must be lined with enough cheesecloth to allow you to fold the ends over the top of the cheese.

Slice off two tops and ladle in the curd in thin slices. When it has drained a little, slide on the tops.

After an hour or so—the time is not critical—pull the muslin up and inwards to curl the outside edges of the curd, a characteristic of this cheese. Do this another three or four times during the next 4 hours.

After draining for up to 36 hours—certainly a minimum of 24 hours—the cheese will be firm enough to handle. Remove the mold and peel off the muslin carefully.

The cheese is eaten straight away, usually as a sweet. The hollow on the top lends itself admirably to being filled with fruit and whipped cream. It is not traditional to salt it, but there is no reason why you should not make it into something savoury.

Cambridge or York cheese

This famous cheese is still sometimes sold. The proper mold is about 8 in (20 cm) long, 5 in (12.5 cm) wide and 6 in (15 cm) deep. Alternatives include square or rectangular cake tins of the same capacity, with the bottom cut out.

This recipe makes one cheese in such a container. The cheese should have an orange stripe through the middle, but it is better to forget this than to use anything other than annatto for the colour.

$1\frac{1}{2}$ gallons (7 litres) milk
$\frac{1}{2}$ fl oz (14 ml) buttermilk
56 drops rennet
56 drops annatto

Start by heating milk to 90°F/32°C, 85°F/29°C for goat's milk. Add buttermilk and rennet diluted with six times its volume of cold boiled water. Stir for 2 minutes then remove 4 pints (2·4 litres) and colour it with annatto.

Do not top-stir the remaining milk as this time you want the cream to rise. Cover both containers and leave for 1-$1\frac{1}{2}$ hours or until the finger test is successful.

Now press your mold on to the creamy uncoloured curd and slice off the top.

Put the mold on a straw mat or cheesecloth on a drainage board and carefully fill with slices of curd. Put in two-thirds of the uncoloured curd, then all the orange curd, then the rest of the plain. This should get the stripe in the middle of the drained cheese. Finish with the creamy top.

The curd should drain for up to three days without turning or salting; eat as is.

French cream cheese

1 gallon (4·5 litres) milk
1 quart (1·2 litres) cream
$\frac{1}{2}$ fl oz (14 ml) buttermilk
15 drops rennet

Mix the milk with cream and buttermilk. Heat slowly to 60°-65°F/16°-18°C. Add rennet diluted in six times its volume of cold boiled water. Stir for 4 minutes then top-stir for a minute.

Put the container, covered, in a safe cool place and top-stir for 2 minutes twice during the next hour. It will be at least 12 hours before you can break the curd cleanly.

Ladle the curd into a cheesecloth, tie up the cloth by its four corners, and hang in a draught-free place until the whey is only dripping slowly. Every hour, scrape the dry cheese from the cloth and mix it back. Once the cheese is good and firm—after about 12 hours—salt to taste and pack into small molds of about $2\frac{1}{2}$ in (6 cm) diameter on drainage mats.

As soon as the cheeses have taken the shape of the molds, they may be removed. They are ready for immediate consumption or may be stored in waxed paper or aluminium foil for up to a week.

Alternatively, use half as much rennet and leave to coagulate for 24 hours. Let the cheese drain on a board rather than hanging it and mold it in much larger molds.

For a Neufchâtel variation, rennet at 86°F/30°C with only a couple of drops of rennet to bring the curd in 18-24 hours. This cheese is molded in a number of ways including the ubiquitous heart.

PRESSED CHEESES

These cheeses require a lot of milk, a press and larger, stronger molds. You need to be certain of your interest and ability before lashing out on such items.

If you have practised on soft cheeses you will have enough grasp of technique to cope with semi-hard cheeses.

Kingston cheese

This is a coloured pressed cheese found in a handwritten cheese-making book marked "Swindon, 1916". It is a Cheshire type of cheese, but with a shorter ripening time of about two weeks, which makes it of interest to the smallholder.

6 gallons (27 litres) milk
$\frac{1}{4}$ pint (·12 litre) strained cultured buttermilk
80 drops annatto
150 drops rennet
salt

To the whole milk add the strained cultured buttermilk and the annatto, diluted with water. Let rest until a 25-second acid test (·19%) is reached. To raise the acidity you may gently warm the milk to 84°F/28°C and hold it there.

Add rennet diluted with five times its volume of cold boiled water.

Deep-stir for 5 minutes, then top-stir until coagulation begins. Cover and leave until the curd is firm. Cut the curd into 1-in (2·5-cm) cubes. Acid should now be ·12—·13%. Gently stir the curd for 15 minutes, then take 20 minutes to raise the temperature to 90°F/32°C, stirring all the time. Continue until the curd is really firm, and the pieces do not stick together. Still at 90°F/32°C leave the curd to pitch (settle), which will take up to $1\frac{1}{4}$ hours. Then leave it in the whey for another 20 minutes. Whey acidity should be ·15-·16%.

Run off the whey and pile the curd in a heap, covered, for 15 minutes.

Cut the curd into 4-in (10-cm) cubes, turn and cover for another 15 minutes (cheddaring). Do this two or three times. Leave yet another hour then grind the curd very fine and add 1 oz (28 g) salt to every 3 lb (1·2 kg) curd. Put into truckle molds.

Press for 2 hours with just enough pressure to compact the curd. Turn and repeat. Turn and leave overnight, without pressure. Leave the cheese in the molds until a slight greasiness appears—about 24 hours. Next day remove the molds, scrape the cheese smooth, bandage with calico and flour paste. Ripen, turning every day, in a cool but not cold room, for up to 10 days.

Caerphilly-style cheese
2 gallons (9 litres) milk
$1\frac{1}{2}$ fl oz (42 ml) buttermilk
56 drops rennet

Add buttermilk or starter at 90°F/32°C. Leave for 30 minutes; add diluted rennet. Deep-stir for 30 seconds, then top-stir to coagulation. Leave for 45 minutes+. Cut into $\frac{1}{2}$-in (1-cm) cubes, turn and stir. Raise quickly to 92°F/33°C. Hold, stirring for 40 minutes. Let pitch, pour off whey, pile for 5 minutes, let sit, cut and pile after 10 minutes. Break into walnut-sized pieces, weigh salt at rate of 1 oz (28 g) to 4 lb (1·6 kg) curd. Put a damp cloth into a mold, fill, press with 4 lb (1·6 kg) for 10 minutes, turn and repeat twice. Leave for 14-16 hours. Dry, lard and bandage. Turn daily for 14 days.

Backyard cheese
2 gallons (9 litres) cow's milk
$1\frac{1}{2}$ fl oz (42 ml) buttermilk or starter
56 drops rennet
salt

Add cultured buttermilk to milk at 90°F/32°C (85°F/29°C for goat or sheep milk). Deep-stir for 5 minutes, rest for 45 minutes. At same temperature, add diluted rennet, deep-stir, then top-stir until coagulation—about 15 minutes. Remove from heat. After 45 minutes, gently cut into $\frac{1}{2}$-in (1-cm) cubes; stir for 5 minutes. Take 30 minutes to heat to 100°F/38°C (95°F/35°C), stirring constantly. Hold firm until curd does not stick together when squeezed, drain, tie in cloth, rest for 1 hour, tightening from time to time. Break up, mix in $\frac{1}{2}$ oz (14 g) salt. Line mold with cloth, pack in curd neatly, fold cloth over, press at about 25-30 lbs (10-12 kg) for 2 hours. Rewrap. Double pressure overnight. Rewrap. Again double pressure for 24 hours. Coat with lard and muslin. Mature for 6-8 weeks, turning daily for 3 weeks then slowly reducing to weekly.

GOAT- AND SHEEP-MILK CHEESES

In many ways these cheeses are much simpler to make than the others. It is easier to keep a goat or sheep than a cow, and their milk does not usually need pasteurizing, although that is a good safety measure.

Lactic goat- and sheep-milk cheese

Lactic cheese, that which curds naturally, is made by standing the raw milk at a temperature of 75°-85°F/24°-29°C for up to 24 hours, draining in something with holes in it and finishing it under moderate pressure. If you wish to aid ripening, add some starter to help—4 fl oz (114 ml) buttermilk per gallon (4·5 litres) milk is about right. If you add a starter, your cheese should be ready in just under 20 hours.

To avoid making very soft curd that is difficult to strain, cheat by adding rennet at the rate of 1 eggspoon per gallon (4·5 litres) milk heated to about 75°F/24°C. The longer you take to set the milk the smoother will be the eventual cheese; with rennet you should aim at 5-10 hours. Milk that is renneted should be about 12 hours old. Flavourings for this and Chèvre paysan are the same as those on page 41, including the addition of cow's cream to the curd.

▲ Goat's-milk cheese stacked in the open air for drying, but shaded from the full glare of the sun.

Chèvre paysan

2 gallons (9 litres) cow's milk

1¾ pints (1 litre) whey

Into the tepid milk stir whey left from another cheese. Leave in a temperate spot for the curd to firm—up to a full day even on a moderately warm summer's day.

Sterilize and dry some small molds with pierced bottoms (which you will probably have had to make yourself). Stand them on a wire rack or on straw and ladle the curd in. Slice, and distribute the slices evenly, i.e. do not fill one mold at a time. Let drain about 48 hours, unmold, shape with your hands and dry in a cool shaded room.

Banon

This cheese can be made with either goat's milk or a sheep/goat mixture. Follow the recipe for lactic goat cheese, but leave the molds to drain on straw or unglazed earthenware—on bricks, perhaps. The cheeses should be rather dry and need 3-5 days to reach this state. Then unmold, plunge them into some *eau de vie* or a rough liquor such as *marc.* Wrap them in chestnut leaves and secure with raffia or string. The curing will continue, the paste turning much creamier and becoming perfumed by the tannin in the leaves. Nettle leaves are also excellent for curing this kind of cheese, although it will not be a true Banon.

WHEY CHEESE

Gently heat the whey until it is simmering, then keep it at that temperature until all the moisture has evaporated and you are left with a paste that will have caramelized. Cool the saucepan in cold running water, stirring the cheese to cool it evenly. It should end up a medium golden brown and have a fairly firm consistency. Press it gently in a fat square mold (a tea tin is fine).

ALBUMEN CHEESE

This is made two ways, either by reaping what you may from the unimproved whey or by adding extra milk—the method used for Ricotta.

Method 1 Heat the whey until it is foaming. Scoop out the small curds, drain in muslin, salt and press slightly.

Method 2 When the whey is foaming, add up to twice its volume of fresh milk.

The liquid must not boil, and when you see the curds begin to separate, remove from the heat, cover and leave until the curd has pitched—about 2 hours.

Drain for about 3 hours in muslin. Salt and pack into a pretty mold.

Buying and storing

Buying cheese really *is* putting your money where your mouth is. Although much cheaper than meat as far as nutritional value goes, it is no longer a cheap food. The demise of the corner shop and the blossoming of the supermarket have also robbed the customer of the chance to taste and discuss. Experience is the only infallible teacher. But here are a few observations that will help you on your way to better buying.

In an ideal world, cheese would reach the consumer in perfect condition, thoughtful retailers refusing to sell anything underripe, overripe or otherwise imperfect. But if you see a cheese that does not look quite right, it does not necessarily mean the retailer has been careless.

You and your shop

A small shopkeeper may be trying to provide a specialist service, building up a trade in traditional and interesting cheeses. Invariably some will sell faster than others, so some may not be in tip-top condition. They are probably perfectly acceptable to many people, sometimes through preference, sometimes through ignorance. They are still worthwhile as a way to learn a little more about cheese — putrefaction and desiccation always excepted. Complain, but don't give up.

The only sure way to help avoid this problem and dissuade your shopkeeper from taking the easy way out—selling pre-packaged basics —is to buy more cheese. When you try something new that you like, tell your friends where you bought it. Try tasting before you buy, so that you can avoid anything that is not quite your style.

In supermarkets the problems are very different and very difficult. Supermarket counter-staff need know very little about cheese to display it neatly and price it properly. If you buy something really bad from such a source, the wrapping having prevented accurate assessment, complain. Otherwise

▶ London's most famous purveyor of fine English and Continental cheese was established in 1797. The families involved had already had over one hundred years' experience of cheese-selling in London's bustling markets.

These three centuries of expert knowledge and tradition are recognizable the moment you step into the sawdust-floored store in fashionable Jermyn Street: refrigeration is eschewed and you may taste every cheese before you buy.

the cheese-buyer may never know what has gone wrong. It may not even be his fault; his supplier may be sending inferior goods.

Buying blind

Buying an unusual or new cheese is full of pitfalls, even for the initiated. There are no real short-cuts — personal experience is the best teacher. You can take out some insurance by going to a reputable shop that still cuts from whole cheeses. If these are good, the rarer, smaller delicacies should be too. What you are looking for is probably soft and French. Remember that these are far more difficult to keep than pressed, domestic cheeses. Once ripe they usually have a very short life. Thus, although some of the grander shops flout modern practice and keep some of these on their counter-tops rather than under refrigeration, this is a matter of display rather than good husbandry. You might be lucky to find something in just the right condition. But often cheese from such a display will be inedible, and because it is new to you, you will not know that it is inedible.

Genuine interest in cheese requires application, thought and planning. As with wine, attention to correct storage, temperature, maturity and accompaniments will pay handsomely.

The subjectiveness of taste buds means no one can give simple rules for your guidance through the forest of trade names and pre-packaged cheeses. Some are certainly better than others. If you really believe the brand your retailer stocks is inferior to another you know, show him the label, hoping it also reveals the supplier's name. Your shopkeeper will love you for it if you are right; still be interested if he thinks you are wrong. As most small shopkeepers cannot find time to try cheeses which they do not carry as part of their own stock, any specialist help you can give is bound to be appreciated.

Generally, the rules applying to the purchase of most cheeses are very much a matter of good sense and similar to those for other fresh foods: there should be no dryness, cracks, sweating, oiliness, mould in the wrong place or of the wrong type, no sourness, decomposition or excessive crumbling.

A buyer should never reject a cheese simply because it has rind (like Farmhouse Cheddar) or because some foil is included in your weighed portion (as in Dolcelatte). It is fair enough to object, however, if inexpert cutting means there is a *disproportionate* amount of rind or foil.

Pre-packed cheese should always be checked for date. The packing should never be smeared or cloudy; avoid any that are puffed up like cushions, even if they are in date. Instructions for opening, storing and serving packaged cheese should be strictly observed.

Vegetarian cheese

No—it's not a joke—nor is it made with dandelion milk! Killing or mistreatment of animals strikes many people as unnecessary, and strict vegetarians usually abstain from cheese because of the use of animal rennet. Observance of Jewish dietary laws presents a similar problem; for they forbid the mixing of milk and meat products.

So what is the solution? The juice of plants is used to curdle the milk. It is a technique known for thousands of years and may even pre-date the use of rennet. A number of wild flowers can serve as curdling agents. Ladies' bedstraw, also known by the country name of "cheese rennet", has been used in England for hundreds of years. The ancient Romans used thistle, safflower and fig sap, which is used today by the Bedouins of North Africa. In Portugal's southern mountains the leaves and flowers of a wild thistle are used: Spain's Cordoba also makes a vegetable rennet cheese.

Perhaps the most amazing coagulant ever used was by the monks of the vegetarian Cistercian order in medieval Denmark —they used the digestive juices of the Venus flytrap and sundew.

Storing cheese

The basic rule about storing pieces of cheese is—don't! Buy in small amounts that you can eat quickly. If this is impossible for you, simply remember that successful storage depends on avoiding the effect of heat, air and desiccation.

It is in fact untrue to say that cheese should never be refrigerated. Of course, a cellar or larder at 50°–60°F/ 10°–15°C is best. Otherwise the vegetable container at the bottom of a refrigerator is perfect. In either case, take note of the following:

Soft cheeses Always keep wrapped in cling film or aluminium foil. The same applies to goat cheeses.

Pressed cheeses It is well worth the trouble to wrap these in a cloth moistened with water, light brine or a mild vinegar solution (plain water is easiest and perfectly efficient).

Gruyère and Parmesan Damp wrapping is good for these and all other cooking and grating cheeses.

Blue cheese This may be closely wrapped in film or foil, or in a damp cloth; the former method is better.

It is worth noting that a crumbly surface dries faster than a smooth one—draw the flat of a knife over the surface of crumbly cheese before storing it.

Fortunately, it is fairly difficult to ruin cheeses totally. Sweated fat can be wiped off, mould can be scraped off—neither is a health risk. Dried cheese can be grated and used as a condiment or cooking ingredient; the simplest ploy with dried blue cheese is to mix it up with butter into a savoury spread. Find a better storage place or method next time and no one will ever know!

For those blessed with neither refrigerator nor cellar, a breezy window ledge or stairwell are good summer substitutes for storing cheese.

What must be avoided at all costs is the cheese dish or cheese bell—a murderer of cheese and incubator for all sorts of undesirable elements.

Above is a choice which will confront the buyer more and more—two types of Camembert. On the right the more common, creamery-made pasteurized cheese—excellent when just ripe, like this example. Left is a Camembert made with raw milk—*"garanti non pasteurisé"*—offering fuller flavour and, of course, authenticity, even if not a true "fermier" (made in a farmhouse).

Any farmhouse cheese or cheese made with raw as opposed to pasteurized milk will cost you more—up to a third extra. Experts can tell you the exact farm or area from which a raw-milk cheese came, for the combination of locality-specific grasses, herbs and flowers are detectable in the flavour. This classification is impossible with pasteurized milk.

▲ A choice of over fifty cheeses, cut to order, with tastings and personal service! Such shops still exist throughout Europe if you look hard enough. This is Mr Christian's, just off London's famous Portobello Road Antique Market.

▼ M. Androuet is a great authority on French cheese and author of the *Guide du Fromage*. He uses his knowledge practically; here in his Paris shop he sells a wide range of authentic and rare cheeses from all over the country.

Literary morsels

Thou hast poured me out as milk and curdled me as cheese.
The Book of Job

I did greedily put in my mouth a great morsel of barley fried with cheese. It stuck so fast, being soft and doughy, in the passage of my throat, that I was well nigh choked.
Lucius Apuleius, *The Golden Ass* (*c.* AD 150)

We have ripe apples, mealy chestnuts and a wealth of pressed cheese . . . there are little cheeses too, dried in a basket of rushes.
Virgil, *The Georgics* (*c.* 30 BC)

Cheese is the biscuit of drunkards.
Grimod de la Raynière, *Almanach des Gourmands*

A dessert without cheese is like a pretty woman wanting one eye.
Brillat-Savarin, *Aphorismes du Professeur*

How can one possibly govern a country [France] which produces over 370 cheeses?
Attributed to both Charles de Gaulle and Winston Churchill

Serving cheese

Full enjoyment of even the most unassuming and inexpensive cheese can be ensured with just a little care; expensive, rare and delicate cheeses demand it by right. As with wine, temperature, atmospheric conditions and the method of serving can have dire or divine effects. Much of the following is simply good sense. Nonetheless, any knowledge you can acquire concerning cheese will help you to ensure that your cheese is served in peak condition.

Serving temperature

This is the most important factor in ensuring you enjoy the full flavour of your chosen cheese.

A moderate room temperature is best for most cheeses. When removing cheese from the refrigerator, take into account the effect of central heating in winter or cool kitchen breezes in summer. An average of half to one hour to reach the right temperature is usual.

On balance, it is probably better to serve cheese slightly too cool rather than too warm — warmth leads to sweating and unpleasant taste changes. To avoid the effect of air deterioration and prevent exchange of flavours, do not unwrap the cheese until just before serving.

Cream cheeses are usually best chilled. But as this convention is largely to safeguard freshness, you may enjoy the fuller flavour of a warmer cream cheese, provided you are confident of its wholesomeness.

The art of presentation

The appearance of the cheese you serve is a matter both of good sense and a sense of respect for your choice.

Trim any dry, cracked or mouldy parts if necessary.

Never camouflage the cheese with gardens of salad, forests of pine or crêches of straw. Let the cheese speak for itself on a plain base. A wooden cheese board is traditional; but if you have a large one and want to serve only one or two cheeses, improvise with a small contrasting plate, a bread board or table mat.

The common practice of covering boards or plates for cheese with split-cane table mats is not recommended. Crumbs of cheese and other food can make them very unhygienic; in any case they are difficult to cut on and serve from.

A few choice condiments may be served with cheese for those who require them. This is more likely to be done

◀ Knowing the correct way to cut cheese enhances its appearance and saves you money.

when cheese features as a large part of the meal—say, a light lunch or picnic. Then fruit, nuts, cress, fresh herbs, chutney or pickles may be offered, with regard to complementing rather than masking the flavours of the cheese.

Pickles—especially pickled onions, traditionally served with a "ploughman's lunch" in pubs—dramatize the astonishing tastelessness of much modern Cheddar. They overwhelm what little *goût* there is—which, perhaps, is no bad thing. But with some well-matured, summer-milk, farmhouse cheese a pickle or two can be ideal.

Wraps, rinds and tags

Depending on the formality of the meal, any wrappings, boxes, etc. should probably be removed; so should inedible rinds.

Some experts maintain no rind is edible, including those of Brie and Camembert. Others allow the possibility of eating the rind of any of the soft paste cheeses with washed or bloomy natural rinds. Use your own taste as guide—making sure that if you do serve the rind, it is not dry, ammoniacal or otherwise contaminated.

The idea of putting name-tags on cheese at the dinner table is appalling. If you are unable to remember the names of the cheeses you are presenting, you are probably serving too many. Unfussy presentation of an interesting or rare cheese will create far more of a sensation.

Tags are unhesitatingly recommended in catering establishments with inexperienced staff and at parties specially held to experiment with, or introduce, new cheeses.

To cut or to scoop?

Unless you are serving children or the infirm, never pre-cut cheese into portions; it hastens the deterioration of both flavour and friendship!

But as there are guests who seem to possess an unerring talent for mutilating cheese, there is nothing wrong in actually serving the cheese to your guests yourself. A pile of plates, a sharp knife (two if blue cheese is included) and a nice line in chat will ensure no one feels you are being mean.

Butter should always be available, though its use is a personal matter. A choice of biscuits—some salty, some

▶ This Swiss speciality is not one but several cheeses; Raclette is a kind of family name. All have special melting qualities which have never been equalled elsewhere. For authenticity, you should rest a great chunk of a Raclette cheese in front of an open fire. When the surface has melted, quickly scrape it on to a plate with potato and gherkins. Especially popular as an *après-ski* treat, it is equally good served outdoors in summer.

not—is best at dinner; bread is better at lunch and for snacks. As all cheese is salted to some degree, never offer just salted biscuits. If you are able to serve only one sort of biscuit, choose water crackers or bath olivers, with salted butter. Thumbscrews and the rack are an inadequate response to those who serve cheese-flavoured biscuits.

▲ Some ways to waste your money: examples of bad cheese! From top left, clockwise: Pont l'Evêque type—too runny, cracked rind with white mould; Brie with chalky centre (too young); Brie, liquid, cracked, dry rind (too old); Cheddar, desiccated and oily; Camembert, brown rind indicating excessive age; Tomme au Raisin, mould amongst the coating, browning under it; Pipo Crèm, excessive mould and browning of body; Stilton, lack of bluing right to the skin indicates immaturity.

What about scooping? In a word, never. Although the centre of a properly aged Stilton is certainly the bluest part, it is no riper than any other section. The flavour simply changes as you move from the rind to the centre, as the amount of blue increases. But no part is in any way lesser Stilton; it all shares in the fascinating spectrum of tastes presented by a mature cheese.

Defenders of the practice are incorrect to say they are logically following the process of ripening from the centre to the rind. Cheese ripens *anaerobically*—without direct air contact. Once air gets at cheese it begins to activate spoilage bacteria. Stilton scooped from the centre slowly changes in texture—certainly. But this change is the result of deterioration and desiccation.

Equally frowned upon is the alcoholic dosing of cheese—again inflicted most often on the noble Stilton. This was originally done, not for a *recherché* taste thrill, but to keep whole cheeses moist.

The party's over

As soon as is decent, get the cheese off the table and back into its protective wrapping. Cigarette smoke and a warm atmosphere are hardly a cheese's best friend. As well as protecting flavour, quick action will mean fewer dried edges to trim and therefore will save you money, too!

Cheese and wine

Cheese is incomparably gregarious. It goes well with great and modest vintages and with beer, ale, cider, fruit, herbs and nuts. Brighten up your diet by making the most of these delectable partnerships.

▼ The beverages: (from left) 1 farmhouse cider; 2 traditional ale; 3 vintage port; 4 Rhône red (glass); 5 Australian Hermitage; 6 sweet white Bordeaux; 7 light Burgundy; 8 Provençale white (glass); 9 Champagne (glass); 10 Alsatian white; 11 rosé; 12 non-vintage claret (decanted); 13 dry madeira; 14 vintage claret.

The cheeses: a Farmhouse Cheddar; b Blue Stilton; c Roquefort; d Chèvre log; e Brie; f unpasteurized Camembert; g Pont l'Evêque; h Farmhouse Caerphilly; i Red Leicester.

Combinations of cheese and beverage: a with 1, 2 and red wines; b with 3, 6, 12 and 14; c with 4, 5, 9 and 14; d with 7, 8, 11, 10 and most light wines; e with 7, 12 and 14; f with the best of 7, 14 and 3; g with 5, 7 and 14; h with 9, 10, 11, 13 and most dry whites; i with all clarets and 13.

There is probably more snobbish rubbish written about the companionship of wine and cheese than about any other food: this cheese must be eaten only with that wine, that with this—the list of arranged marriages is endless. Preferences are subjective in the extreme; one sometimes suspects the unsubtle marketing scheme of an ambitious wine merchant.

A guide without limits

The truth is that very few of us, even the most devoted sybarite, will have the cash

or the inclination to go out and buy a highly individual bottle of wine just to complement a cheese. The wine is usually there first—bought to go with the main course or, more likely, left behind after the last party. Cheese is expected to slot in.

The joy of both wine and cheese is that neither has tiny, cramped, limited spheres of conviviality. Whether great or simple, they can adapt to their surroundings.

Broadly speaking, the cheeses and wines of an area do have a natural affinity, particularly if you are drinking the simpler wines. Finer vintages spread their welcoming arms wider. Experience teaches you that this is but a basis for pleasure. For a light claret easily replaces a Beaujolais and a really crisp white wine can more than equal the light reds, and so on. But overripe Brie will never be saved by a superb Bordeaux or Burgundy. Neither will an absolutely perfect cheese salvage or camouflage a sour or vinegary wine. The general rule is to ally full-bodied wines with richer cheeses; the lighter white and rosé wines deserve the fresher and creamier cheeses and the subtlety of light, newly-made Chèvres.

Cheese and wine at the dinner table

Your choice of cheese will depend on when it is to be served. If this is to be between the main course and the pudding—as the French do—it is usual to have it matched to the main-course wine. If you had served a white Meursault, it would be better not to offer a Brie or a tangy Blue Cheshire to eat with the last of that wine. A soft goat's cheese, some flowery Gruyère or a Tomme would be much more balanced choices.

If you have offered a particularly good main-course wine, you should always finish it by serving cheese before your pudding.

This is even more important if you are to serve a luscious, sweet, chilled pudding wine. You will thereby ensure full appreciation of both wines and cheese and avoid the horrid taste of red wine with a sweet pudding.

On the other hand, any diner smart enough to change wines is likely to finish with port or madeira, in which case cheese is also a highly suitable accompaniment; but on balance it would still be better to serve it before the pudding.

Those who insist on finishing with cheese, and who have served a red wine, might consider avoiding pudding altogether; if they wish both to finish with a sweet pudding wine *and* cheese, such strongly flavoured cheeses as Roquefort make for a sensational combination—if you can afford them.

Wine, cheese and conversation

Entertaining a special friend or two gives one a chance—and an excuse—to make a special effort. Here you really might find it worthwhile to purchase a specific wine and a particularly interesting cheese. Try a Gaperon from the Auvergne and a flinty wine from Corbières. Or a creamy Fontainebleau and a still Champagne; Taleggio and something unflinchingly Italian such as Vernaccia.

If you feel inventive, cross international borders and team some perfect farmhouse Caerphilly with a chilled, dry Provençale rosé. While the classic combination of Stilton and port is unbeatable, slightly pungent cheeses also complement a sweet wine, like a Sauternes. So perhaps a Chateau d'Yquem with some Roquefort! A cheaper Monbazillac would also do very well. Sherries and madeiras offer wonderful opportunities to experiment.

Cheese and wine parties

It must be said that, as a way of entertaining, cheese and wine parties are a bit dated. At their zenith, in the 'sixties, excitement was generated largely because of the growing variety of cheese and wine available. Now inclusive holidays, the media and better-stocked shops have made us more sophisticated.

There is still no simpler or more appreciated way of catering for your guests than with an excellent cheese board. It is not a bad idea to arrange the cheeses according to their strength—even to separate them—having those more suited to the white wine close to it, and so forth.

Unless you are a wine merchant wishing to show off some new wines or are determined to explore the gamut of taste experience in one go, stick to one red and one white wine, and some decent-sized chunks of a limited number of cheeses. Better a pound or two of a couple of good cheeses than bits of eight or nine. Imagine what the latter will look like when the guests have started in on them!

Again, unless you have arranged the evening as a "study session" for cheese-counter assistants, do not present guests with carefully labelled cheeses in front of carefully labelled wine. It smacks of regimentation and will ruin your evening the moment you see the first person make the wrong mix! Food, no matter how good or how interesting, should never be the star of your entertaining. That honour should go to you and your guests.

How much cheese should you buy? People eat far less than you think—unless they came expecting dinner. If you keep the selection simple $1\frac{1}{2}$-2 oz (42–56g) per person is more than adequate. If you must have a wide selection, buy a fraction more—allowing for that wasted by people slicing off more than they want from a wider choice than they can actually absorb.

Cheese and other drinks

Wines made at home from fruit, vegetables and herbs make delicious accompaniments to cheese—both the pressed English types and the Continental specialities. Elderberry wine goes down excellently with Stilton, mature Cheddar, and possibly even Blue Cheshire. Gooseberry was recommended by Victorian connoisseurs as a complement for Caerphilly. Those making parsnip, onion or pea pod wine will need to do some personal experimenting, since little literature exists on these.

And of course, there is cheese with ale and beer—a truly great combination and one very popular in many European countries, particularly those with no great red wine tradition. Brie, Camembert and blue cheeses do not go well with beer, but most of the pressed cheeses love it. So do newer English cheeses with chives and spring onions. It is an interesting way to entertain and as you cannot go very wrong, you can serve a variety of beers.

► A highly undesirable marriage of too many cheeses and too many wines in too hot a room, too near people and cigarette smoke.

DUCHENE

Cooking with cheese

Some cheeses can be appreciated in a wide variety of ways. The best cheese cookery will enhance, rather than disguise, a cheese's individual characteristics. However, some types, such as blue cheeses, should hardly ever be used in cooked dishes. This collection of recipes is by no means exhaustive but may serve to introduce you to some less common ways of using cheese.

Cheese and cheese recipes seem to find their way into most cookbooks. Here, recipes have been chosen for their historical interest, flavour and lack of gimmickry. For some respect should surely be granted our more famous cheeses. One restaurant in London offers Camembert and Champagne soup—it may taste good, but does it respect the flavour of the Camembert? There is such a thing as Stilton ice cream, and Roquefort has been suggested as a topping for cauliflower cheese. If you wish to try these combinations, it is your prerogative; but the less ambitious—and less "original"—combinations are usually more successful and satisfying.

◀ Round Europe with cheese cookery. From the top, clockwise: Cheese Loaf, Gougère, Quark Tart, Fish Pie, Fondue, Black Cherry Cheesecake, Mozzarella and Orange Salad, Spinach Quiche, Pizza, Potted Cheese, Cheese-stuffed Potato, Coeur à la Crème, Cheese Scones, Macaroni Cheese. Centre: Gouda Cornets, Greek Salad.

Brie tart

Surprisingly, Brie is often served hot, and there seems to be a tradition of some centuries behind this. Croquettes are one familiar way of serving it, rolled in a mustardy sauce, breadcrumbed and deep-fried. The following recipe for Brie tart is from the fifteenth century. Modern Brie is nothing like the cheese of that date, but the tart is delicious and well worth trying.

8-in (20-cm) blind-baked pastry-crust base
1 lb (·4 kg) Brie—just ripe
4 fl oz (114 ml) single cream
3 lightly beaten eggs
1 oz (28 g) brown sugar
pinch powdered ginger
pinch saffron
salt to taste

Put the saffron in a cup and just cover with boiling water to draw out colour and flavour.

Remove and chop the Brie rind, and sprinkle over pastry base. Mix the Brie, cream, eggs, sugar, ginger and saffron-water until smooth. Add the salt and adjust other spices to taste. Bake 30-40 minutes at 350°F/180°C/Gas 4 until set and lightly browned. Serve warm.

Egg cheese

Below is a recipe from Edinburgh dated 1791. Though the ingredients may seem extravagant, the result is worth the effort.

3 pints (1·7 litres) full-cream milk (or milk and cream)
½ pint (·25 litre) sweet white wine
pinch finely grated lemon peel
pinch sugar and cinnamon
12 egg yolks
6 egg whites
mold with drainage holes

Flavour the full-cream milk (or half milk, half cream) with the cinnamon, sugar and finely grated lemon peel. Add the white wine. Beat the egg yolks and the whites together in a bowl and add to the mixture.

Heat very gently until it boils and sets. Turn into a fancy mold which has draining holes (finding such a mold is probably the hardest part of the recipe, but in fact an ordinary sieve will do). Let stand until the whey runs out—usually overnight.

Chill before serving with orange or rose water sprinkled lightly over it. Decorative and flavourful additions include fresh berries or mandarin oranges.

Cheese loaf

A simple name for a rich, satisfying dessert that represents late-eighteenth-century cooking at its most hearty and attractive.

3 pints (1·7 litres) fresh full-cream milk
fresh breadcrumbs
12 egg yolks
6 egg whites
pinch cinnamon, nutmeg and sugar
10 fl oz (284 ml) double cream
glass brandy
salt
1½ oz (42 g) flour

Curdle the full-cream milk and gently press out the whey. (Alternatively buy fresh full-cream cheese.) Add an equal weight of breadcrumbs to the resulting curd. Beat the egg yolks and whites together and add to the mixture. Stir in the cinnamon, nutmeg, sugar, cream and brandy.

Knead this dough until smooth and satiny, adding a little salt and the flour to counteract any stickiness. Shape into a loaf and bake at 350°F/180°C/Gas 4.

When still hot, cut a slice off the top and pour in a sauce made to your taste of melted butter, cream, sugar and brandy.

Gougère

This is a nineteenth-century speciality from Burgundy. It can be all sorts of things—a hot hors d'oeuvre *par excellence*, a tangy accompaniment to game and poultry or a cake for any time!

(1) Hors d'oeuvre

6 fl oz (170 ml) water
2 oz (57 g) butter
pinch salt
6 oz (170 g) flour
3 eggs
3 oz (85 g) Gruyère

Boil the water, with the butter and the salt. Sprinkle in the flour, then stir well and cook slowly for about 30 minutes, or until the taste of flour has disappeared. Cool slightly, then break in the eggs, one at a time. Do not add another until the preceding egg has been well beaten in. Still stirring, add the Gruyère, cut into small cubes—it must *not* be sliced or grated. Make a ring of the pastry on a greased baking tray and bake for 20-30 minutes in a hot oven: 400°F/200°C/Gas 6. *Do not* open the oven door until the last possible minute, or the pastry will collapse.

(2) Cake

½ pint (·25 litre) water
4 oz (114 g) butter
pinch salt
8 oz (227 g) flour
5 eggs
8 oz (227 g) Gruyère

Boil the water, butter and salt together. Remove from the heat and add the flour. Stir in with a wooden spoon, and replace on the fire, stirring continuously. Cook until almost dry. Remove from the fire. Beat in the eggs, one at a time. Chop finely—*do not* grate—the Gruyère into the paste and mix in thoroughly. Save some of the cheese to sprinkle on top.

Place the mixture one tablespoon at a time on a greased baking sheet, forming a ring. Brush with beaten egg, sprinkle with the remaining cheese and bake in a medium, 350°F/180°C/Gas 4, oven. Enjoy it when cold with a red or white Burgundy.

Spinach and cheese pie

Spectacular to look at, spectacular to eat.

2 lb (·9 kg) spinach leaves
salt
1 oz (28 g) chopped parsley
1 oz (28 g) chopped fresh dill weed (less if dried dill weed is used)
4 eggs
8 oz (224 g) Fetta cheese (or white Stilton)
pepper, salt, nutmeg
***Phyllo* pastry**
melted butter or olive oil

Wash the spinach and drain or sponge it fairly dry. Cut it up roughly and sprinkle with a generous amount of salt. Place in a colander to drain for several hours. Squeeze dry. (This technique for incorporating spinach in pies and quiches prevents the problem of sogginess.) Mix the beaten eggs, herbs, pepper and extra salt needed, and add with the nutmeg into the spinach. Toss in the cheese, cut into cubes.

Brush a baking pan about 11 x 7 in (27 x 17 cm) with melted butter or olive oil. Line with sheets of *Phyllo* pastry (obtainable in delicatessens and Greek shops), brushing each one before putting another on top. Arrange the sheets so there is about 2 in (5 cm) of pastry sticking up over the sides of the dish. Spread the filling evenly and cover with more *Phyllo*, again brushing each sheet with melted butter or olive oil.

Interweave the first few layers with the extra pastry from the sides, to seal the pie. Eight to twelve layers altogether is about right. Trim any excess unsealed pastry from around the edges, and then cut through the top three or four layers with a sharp knife making a diamond pattern. Dampen the edges with water to avoid burning.

Cook at about 350°F/180°C/Gas 4 for at least 40 minutes, sprinkling the edges with water whenever they seem in danger of burning. The top layers should separate and curl, looking as if they are about to float away. Serve as the Greeks serve all their food—warm rather than hot.

Frozen or fresh strudel leaves will adequately replace *Phyllo* pastry. Otherwise roll puff pastry very thin. It will still be a refreshing change from the soggy spinach pies usually encountered.

CHEESECAKES

Cheesecakes hot, cheesecakes cold, cheesecakes sweet and savoury, large and small—all have been popular since the days of the Roman Empire. They were a feature of the banquets of the Middle Ages, always found in mid-European cuisines and, in unbaked form (as descendants of the German Quark tart), have a staggering variety of flavours. Here is a sample of international cheesecake cooking, from the fifteenth century to today's refrigerator specialities.

Fragrant cheesecake

From the court of Richard II.

9-in (22-cm) base crumbled digestive biscuits or a blind-baked shortcrust
3 oz (85 g) dried elderflowers
5 fl oz (142 ml) double cream
4 oz (114 g) castor sugar
2 oz (57 g) fresh white breadcrumbs
6 egg whites

Soak the elderflowers in cream until they are well moistened. Add the sugar, stirring until dissolved. Mash or sieve the cheese until smooth, and add the bread crumbs and then the flower-cream mix.

Beat the egg whites until stiff but not dry; fold lightly into the cheese mixture. Spoon gently into the crust and bake at 375°F/190°C/Gas 5 for 50 minutes, or until set and firm. Turn off the heat and cool in the oven with the door open for at least 15 minutes.

Use fresh elderflowers if you can, ensuring that they are clean and free from contaminants. You can substitute other fresh or dried petals—particularly rose and orange—but always be sure to double the quantity if you are using fresh flowers. Alternatively, use orange-flower or rose water. Ideally, the cake should be served just warm.

Miniature cheesecakes

An eighteenth-century snack, popular as a snack at smart soirées and card parties.

2 quarts (2·25 litres) fresh full-cream milk
junket rennet
6 oz (170 g) unsalted butter
1 oz (28 g) ground almonds
dash orange-flower water
½ glass Madeira, sweet sherry or Malaga wine
4 oz (114 g) currants
dash cinnamon and nutmeg
3 fresh eggs
10 fl oz (284 ml) single or double cream

Turn the rich full-cream

milk with the junket rennet. Cut the curd and strain off the whey. When the curd is dryish, crumble it in your fingers. Mix in the butter and almonds, orange-flower water, the madeira or sherry and the currants. Sprinkle in the nutmeg and cinnamon to taste. Beat in the eggs and cream. Put in patty pans and bake gently for about 40 minutes at 350°F/180°C/ Gas 4. Serve warm or cool.

Käsekuchen

Germany and Austria are famed for their cholesterol-packed baked cheesecakes. Some are as light as a feather, others make it difficult for the diner to rise from the table. But they are never stodgy. This recipe, found in New York, is of German derivation. Unusually, it makes use of cottage cheese, but curd cheese can be used if preferred. Orange juice and rind can be used instead of lemon. Soured cream instead of sweet cream is a nice variation. Crushed cardamom is another imaginative additive.

9-in (22-cm) springform pan with crumbled digestive-biscuit base
4 eggs
6 oz (170 g) castor sugar
pinch salt
juice of 1 medium lemon
grated lemon rind
8 fl oz (227 ml) double cream
1½ lb (680 g) cottage cheese
2 oz (57 g) plain flour
dash vanilla extract

Beat the eggs (room, not fridge, temperature, please), until very thick. Slowly add the sugar, beating until light and smooth. Add the other ingredients, one after the other, continuously beating. Pour gently into the pan, and bake for 1 hour at 350°F/180°C/Gas 4. Turn off the heat, open the door and let the cake cool gradually.

Yoghurt cheesecake

Coffee, chocolate, mandarin, black cherry, Grand Marnier, apple, toffee, mint—you name a topping or a filling and someone has invented a cheesecake to match. This is by no means the simplest refrigerated cheesecake, but its use of yoghurt gives it a topical appeal.

8-in (20-cm) tin with biscuit base
4 rounded teaspoons gelatine powder
5 fl oz (142 ml) hot water
12 oz (340 g) cream cheese
5 fl oz (142 ml) natural yoghurt
3 teaspoons lemon juice
¼ teaspoon salt
10 oz (283 g) castor sugar
10 fl oz (284 ml) double or whipping cream

Dissolve the gelatine in hot water. Beat the cheese, yoghurt and lemon juice until smooth. Combine egg yolks, salt and 6 oz (168 g) sugar. Stir over low heat until thick. Beware of curdling. Add the gelatine solution. When cool, add the cheese mixture. Whisk the egg whites until stiff but not dry, gradually adding the remaining 4 oz (112 g) sugar. Whip the cream. Fold the egg whites and cream into the cheese mixture. Pile quickly into the tin (which already has the crumb base in it) and place in the refrigerator.

For a fruity version, use a carton of fruit yoghurt and use the same fruit to decorate the top of the cake.

All-American cheesecake

In the United States, the biscuit base for cheesecake is always graham crackers—at least according to modern tradition! The American version is also synonymous with Philadelphia cream cheese—a brand name that has become an institution. Of course, these constituents can and must be varied to coincide with local availability—but it doesn't hurt to know the definitive version should you meet it!

¼ lb (113 g) graham crackers or sweet digestive biscuits
2 oz (57 g) melted butter
2 tablespoons cinnamon
2 tablespoons sugar
1 lb 2 oz (504 g) Philadelphia cream cheese
6 oz (170 g) sugar
2½ oz (128 g) sifted flour
pinch salt
1 egg

6 egg yolks
6 fl oz (170 ml) sour cream
1 lemon
6 egg whites
3 oz (85 g) sugar

Crush the graham crackers or digestive biscuits and mix well with the melted butter and cinnamon and sugar. Pack into the bottom of a well-buttered 9-in (22·5-cm) springform pan and put into the refrigerator to firm.

Cream together the cream cheese, sugar, flour and salt. Beat in the whole egg, the sour cream and the juice of a lemon. Beat the 6 egg whites until stiff and then beat in the sugar. Fold into the cheese mixture. Pour into the prepared shell and bake in a moderately hot oven 350°F/180°C/Gas 4 for 1 hour or until set. Open the door slightly and let the cheesecake cool in the oven.

THREE SPICED CHEESES

Like so many dishes of expediency, there are no authentic potted cheese recipes—merely guides to the techniques employed and some successful examples. These are excellent bases for your own creativity.

Potted cheese

Take some left-over cheese, as dry as possible, and grate it. Warm a little butter and brandy, port or wine in a saucepan and mix in the cheese. Heat just enough to keep the mixture soft and watch that it does not separate. Add extra flavourings

that suit your tastes—nutmeg, mace, paprika and freshly ground black pepper are good basics. Others recommended are anchovy essence, tabasco sauce, crushed garlic, chopped parsley, chives, spring onion, celery leaf, dill, cumin, fennel, caraway—or simply more butter and alcohol. Now is the time to crumble in any blue cheese you might wish to add—cooking it is not a good idea. When everything has blended well, pack it into small pots and cover with butter. It keeps for ages in a refrigerator, but if you have included garlic, chives or onion it is better to eat it quite quickly.

As well as herbs you can make it even more interesting by adding chopped nuts—especially walnuts — or chopped ham. If the cheeses are plain, solid and classical, like Cheddar or Gouda, you might like to grate in some orange rind.

It is fun to make a different flavour each time. A small store of potted cheese in the refrigerator means you will, at least, always have a snack for emergencies.

Fromage en pot

This differs from the previous recipe in that it uses fresh cheese. It is basically a homemade product which you will never find in the shops, and is made in various ways all over Europe—in Belgium it is Potte Kees.

Take an amount of the freshest full-cream cheese (make it yourself, buy it, but do not use commercial curd cheese as this has too low a fat content) and hang it to drain in cheesecloth for five days. Take a smallish earthenware crock, fill to about 1 in (2·5 cm) deep, sprinkle with salt, pepper and fennel seeds. Add another layer of cheese and seasonings and continue until the cheese is all used. Seal the crock carefully and set in a cool, dry place for six weeks. When the crock is opened, remove the mould and work the cheese with a wooden spoon. It should be kept well covered. If you decide you like this sort of thing, experiment with tarragon, thyme and bay leaf. You may add brandy, *marc* or vinegar when mashing.

Fromage fort is the name given to this kind of cheese when mature cheeses are used in preparation. *Fromage fort à la lyonnaise* is made with old goat's-milk cheese which is first macerated in leek water. This is certainly an acquired taste, and justly famous for its pungency.

Liptauer cheese

This third type of potted cheese is easier to make than either of the other two. Un-

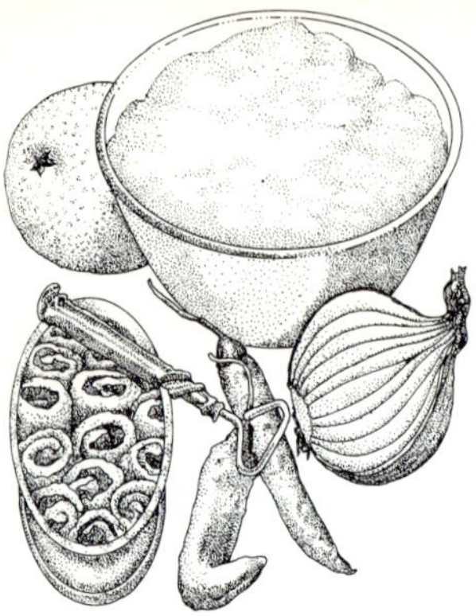

like the English version and the French *fromage en pot,* this Hungarian treat uses unaged and uncooked cheese but is highly spiced.

The original recipe uses Quark, but a good full-cream cheese can be substituted. Into it stir and knead softened butter, a good dollop of anchovy paste, kümmel (caraway) seeds, prepared mustard, a dash of hot or sweet paprika, capers according to taste, salt and pepper and, finally, a handful of chopped onions.

No proportions have been given since, traditionally, every cook finds her/his preferred taste. It should be very spreadable, and almost have a whipped consistency. In Austria and Hungary it is served with rye bread and the light new white wines or lager. There is no more delightful summer snack—but beware, the combination of the tangy cheese and sunshine will make your wine or beer disappear as if by magic!

FINISHING TOUCHES

Any good cookbook will have numerous recipes incorporating cheese. The following notes show how cheese and an enterprising cook can make an inventive contribution to many types of dish.

Hard cheese ends can make super biscuits. Grate them, mix with plain flour and water to a stiff paste, add a dash of salt and cayenne or pepper if you like. Roll thin, cut into strips or circles and bake in a moderate oven until crisp.

A handful of cheese makes scones into cheese scones.

Sauerkraut toast is a heartier use of the bread-and-cheese combination. Top toasted bread with heated sauerkraut, some salt beef, pepper and, finally, thick slices of Emmental cheese. Sprinkle with paprika or caraway seeds before serving. It makes a filling

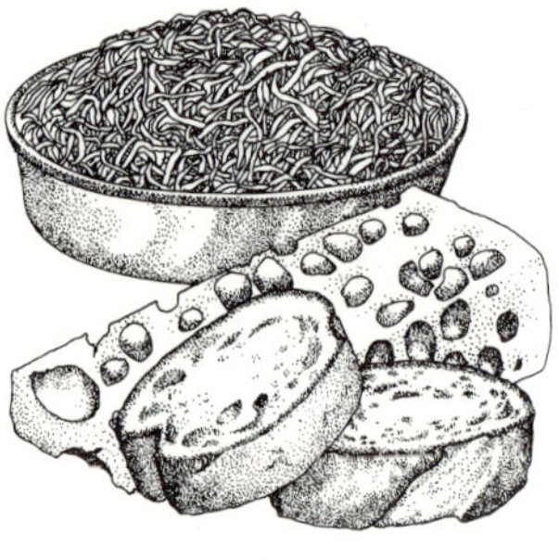

late-night supper or a warming lunch with soup on a winter's day.

For lunch or light supper, split open a baked potato and stuff with grated cheese (Cheddar, Gruyère and Emmental are best), butter and a dab of soured cream or milk. Spice with celery salt, paprika or nutmeg.

Lancashire is the proper cheese for rarebits; add almost anything to the basic mix, from anchovies to ham and pineapple.

Cheese cubes are delicious in salads . . . the Greeks use Fetta; you could use white Cheshire or white Stilton.

If Parmesan is too expensive, look for the ribbed, round, yellow Pecorinos from Sardinia or the Italian mainland. These are wonderful for nibbling, as well as for grating.

Cream cheese is the basis of a luxurious frozen dessert. Flavour with marrons glacés, crystallized fruit and orange or coffee liqueur. That is it! It will not freeze solid.

Cheese and apple do not have to be served separately or cold. Finish cooking a baked apple with grated cheese, or alternate layers of cheese and apple in your next pie.

"Apple pie without some

cheese is like a kiss without a squeeze."

Most vegetables have an affinity with cheese. Excite interest by grating cheese over hot vegetables when you serve them; stuff vegetables such as marrows, courgettes, aubergines and large tomatoes with cheese, breadcrumb and herb mixtures.

Cheese grated into a bowl of soup is a fragrant addition: Gruyère on onion is the great classic; the Dutch also grate Gouda on their pea and ham soups.

The next time you are making a fish pie, mix cheese into the potato topping. For something more substantial, top the pie with a cheese scone mixture but make sure it is well cooked before serving.

Cheese soufflé is always popular and is easier to produce than one imagines. For a nineteenth-century touch, line your soufflé dish with pastry, blind-bake it, then add the soufflé mixture —Queen Victoria approved. If you are making the pastry yourself chop fresh herbs into the mixture for an even more appetizing touch.

A simple party or pre-prandial snack is Stilton-stuffed celery. Break off individual stalks and crisp in ice-water. Leave the leaves on if they are fresh-looking. Cream the Stilton and butter together, and push into the hollow of the stalks. Cream cheese and chive stuffing is also very popular.

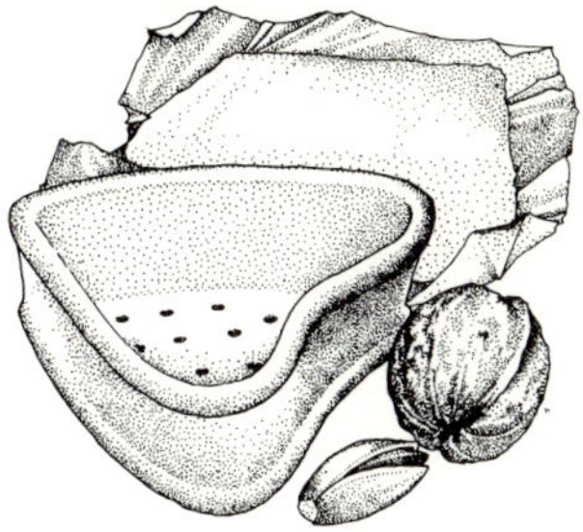

Plain cream cheese can be home-decorated to look and taste every bit as good as the imported varieties. If you are making your own cheese, either drain the renneted full-cream milk over the sink in cheesecloth in the usual way and then pack in a mold to shape—or, if you can find one, use a shaped mold with drainage holes in it. If you use commercial full-cream cheese, just pack it straight into an ordinary pretty mold. Whatever method you choose, you can decorate the result with pistachios, walnuts or pine nuts at a fraction of the commercial price.

There is also a traditional heart-shaped cheese dessert. Just cover the molded cream cheese with fresh strawberries or raspberries in their juice.

Cheese and beer cannot only be mixed and toasted as a rarebit—they also make an easy dip or fondue. About $\frac{1}{8}$-$\frac{1}{4}$ pint (71-142 ml) beer per 1 lb (·4 kg) grated Cheddar with a drop of Worcestershire sauce makes a delicious dip or an easy summer lunch.

► Heat changes cheese flavour and texture. "Stringing" is the worst example, seen most often in Emmental and Mozzarella. Low temperatures, or mixing with wine, can lessen the effect.

Guide to Europe's cheeses

That great cataloguer of mistresses, Casanova, attempted to compile a guide to cheeses during his busy life. It is said to be the only failure to which he readily admitted. Nor have subsequent books succeeded. It is impossible to catalogue *all* the world's cheeses. In any case, there is little point in tracking down local specialities which are unavailable outside the mountain village in which they are produced. This guide introduces a range of cheeses that, thanks to modern means of distribution and refrigeration, is generally available for purchase. Opinions will of course differ concerning which varieties of cheese should be included in such a guide, but it is hoped that a sound basic list has been provided. Note that modern commercial practice often results in the same cheese now being marketed under several different names.

Since this section is a practical guide to buying cheese, the fascinating histories of many cheeses have not been included. Although a comprehensive knowledge of the legends surrounding a particular cheese makes fascinating table talk, it will not help you pick the best of a bad bunch from the cold counter of a suburban supermarket.

In the following guide, each geographical area is introduced by a general discussion of its cheese, followed by a concise, informative list covering the country's best-known or most widely exported varieties. The list describes the type of milk, process by which the cheese is made, fat content, ripening time if known and seasonal availability. Practical information includes how it should look and smell. Where relevant there is advice on cooking and on any other special points which contribute to enjoyment. The following is an explanation of the symbols used:

Milk type

R–raw, P–pasteurized
R/P–either can be used
F–full milk, S–skim
F/S–mixed
An asterisk (*) after an animal's name, e.g. Cow*, means that the more usual animal already noted may sometimes be replaced by another, either as a matter of tradition or availability. In this case, the other animal is identified in the accompanying text.

Cheese type

Explanations of these categories are given on pages 24-29.
1a–cream cheese
1b–acid-curd cheese
1c–rennet-curd fresh cheese
2–soft paste–white unwashed rind
3–soft paste–red/orange washed rind
4a–scalded, lightly pressed
4b–scalded, heavily pressed
5–cooked, heavily pressed
6–plastic-curd
7–whey
8–albumen
9–blue
10–goat and sheep (brackets indicate process)
11–processed

Fat content

This is shown as a percentage, e.g. 45%.

Great care must be taken when considering the fat content of cheese. The percentage given is that of the solid matter in a cheese: 45% fat content does not mean 45% of the cheese you eat is fat; it means that of the solid matter of that cheese, 45% will be fat. Therefore a very wet cheese with 45% fat content can have much less fat per mouthful than a drier harder cheese with a lower stated fat content. Remember, fat content figures alone are no instant guide to a cheese's fattening power. To know that you need much more information. Also, of course, protein is an important element of food value. A simple guide to working out a cheese's approximate calorie count is given on page 90.

Ripening time

This provides an indication of the time it takes the cheese to ripen or ferment to the correct condition. By working backwards you can assess the quality of the milk used; but, sadly, this is of diminishing use today, since it can only be used to assess cheeses made with raw milk. For instance, we know it takes about 4-6 weeks for a Camembert to ripen, although refrigeration may well extend this period. However, if offered an unpasteurized Camembert *Fermier* in February, we could be fairly certain it was made with winter milk and that the cows are unlikely to have fed on lush green pastures. The same cheese offered in June would certainly have been made with rich spring milk. Do not forget when counting back in this way that hot mid-summer can mean very little grass, but that the showers of autumn bring the "second growth" grasses that many believe are finer for cheese-making than those of spring.

Cheeses made with pasteurized milk are fairly similar all year through, though a trained palate might detect slight differences in flavour.

The superiority of summer

▶ Gruyère (top) and Emmental. Both are cooked, hard-pressed cheeses and to the average shopper would look like different sizes of the same cheese. But rind washing and bacterial growth give Gruyère a totally different taste and texture.

milk is the reason why we still regard Stilton as being at its height at Christmas-time. Once this was correct. Stilton takes a good six months to mature. Therefore Stilton made from the first rich raw milk of summer would come into season at Christmas.

Now, pasteurized milk means that Stilton can be of prime quality all year round, and any lapse is usually the fault of insufficient ripening rather than the type of milk used.

In the following lists, the ripening period is indicated in months, e.g. 1m, 6m. Those cheeses which continue to mature beyond their minimum ripening time are indicated by a plus sign (+). Thus "6m+" after "Cheddar" means that though six months is the minimum time needed to produce a fine product, the cheese will continue to ripen after this period, increasing in flavour.

Where a cheese's ripe condition is fugitive and requires refrigeration to keep it wholesome—such as Brie and Camembert—an exclamation mark is added, e.g. Brie 1-2m!

Where the mature version of a cheese is recognized as a different cheese type—as in mature Gouda—a note to this effect will be made.

Season

As a natural extension of the preceding paragraphs, where applicable a cheese's best eating season will be given by indicating the months, e.g. May to July.

P12 indicates pasteurization, and means that the cheese varies little throughout the year.

Characteristics

Now we enter the realm of subjectivity, for the characteristics of a cheese include not only appearance and smell, but also taste. And even if we all privately concur on how a ripe Brie should look, it is almost impossible to express it in words.

What you may call sharp, another may call sour; what has an unacceptable smell to others may be bliss to you. Never take anyone else's word, but fiercely stick to your own opinions—provided you have actually tried the cheese in question.

Variations

Quite apart from the confusion caused by calling the same cheese different names, different cheeses can have the same names! Most of the world's best-known cheeses would once have been made almost exclusively of sheep and goat milk. The explosion of population and easy transportation systems of this century have led to such a great demand for cheese that no herd of the low-yielding sheep and goats could ever cope. So the cheeses were made with cow's milk and a subsequent lack of authenticity. This does not make the new versions inferior; they are simply different, although called by the same name. On the other hand, many cheeses were *always* made with whatever milk was best or easiest at the time. If you are out in the country and are offered a goat's-milk Cheddar, cow's-milk Banon or sheep's-milk Petit Suisse, don't question the validity of the cheese or of this book or imagine ignorance on the maker's part: it's a healthy sign that somewhere the good old days are still with us.

French cheeses

A different cheese every day, and a choice on high days and holidays: that's the range of cheeses made in France. If you are determined enough to nose out the specialities known and eaten only in remote areas, the choice becomes even wider.

Cheeses unlimited

Few people interested in food manage to go long without encountering Brie and Camembert, and probably also know of something called Roquefort which is blue, and certain others which are smelly. This is barely the tip of the cheese mountain. French cheeses run the gamut of taste, from gentle to harsh. They exhaust geometric possibilities in their shapes, can be purest white or coloured and speckled, and vary from rock hard to soft as lightly whipped cream. Some, as they grow slowly from meek infancy to robust maturity, offer such a variety of tastes they are rarely without friends; others bloom just for a day or two—uninteresting before, inedible afterwards. Many of these cheeses are as strictly protected, nationally and internationally, as the famous wines. France is rightly jealous of her heritage.

But just as she allows Brie and Camembert to be made elsewhere (with the proviso that the manufacturing country's name must appear before the words Brie or Camembert) she in return makes the cheeses of other countries. Fetta, Gouda, Cheddar and dozens of other imitations are actually exported back to the countries from which the types originated!

Cultural conservation

How have so many French cheeses managed to survive in an era when "cottage industry" is an exception? Mainly because France never suffered anything comparable to the Industrial Revolution of nineteenth-century Britain, which lured hundreds of thousands of land workers from their rural life. And because the traditional peasant-based rhythm of the French countryside was difficult to up-

▼ Among those present at the Congress of Vienna in 1814-15 were England's Duke of Wellington (seated left) and France's Prince de Talleyrand (seated second from right). It was not all politics and borders. The leaders of Europe found time to vote a Brie de Meaux "King of Cheeses". This was one of the few Congress decisions not subsequently overridden.

set by war. On the contrary, rations and deprivations made farmers and peasants even more determined to stay on the land, which guaranteed them something on the table in return for honest work, a lure with which the towns could never compete.

Each of France's contrasted provinces is a patchwork of self-sufficient micro-cultures. The character of each is determined inescapably by the nature of its soil. Viniculture is the touchstone, for the *vin ordinaire* of an area is largely unalterable, and to change it would be to add unnecessary expense to a very cheap essential. Cheese, on the other hand, is easily adapted to be milder or stronger. The affinity of wine and cheese, both single-ingredient foods based on the flavouring properties of the local soil, is the solid foundation for most provincial cuisines. Where wine is light, the cheese will be too; but when sun and soil produce robust reds and flinty whites, the cheese will be equally strong.

Vigilance and a demand for quality have also played a part in the preservation of gastronomic tradition. While other European nations have seen their farmhouse production decline—and in Britain's case, entire cheese types disappear—the French housewife or *ménagère* has ever been insistent that convenience does not destroy traditional pleasures.

It would be fatuous to pretend that no changes have been made. Pasteurization and commercial techniques have caused a difference. But most of France's millions of small landowners still treasure a milk-producing animal (usually a goat) of some kind, and make their own cheeses.

At the other end of the scale, commercial production of cheese is staggering, considering a vast proportion is directed toward the speciality or gourmet market. Well over 500,000 Camemberts are made every day in creameries alone; this discounts those made traditionally on farms. It is a sobering experience not to be able to buy genuine farmhouse Cheddar in an English market town, but to be offered a choice of Camemberts.

▲ Madame Marie Harel, farm woman of the tiny town of Camembert (Orne). She is credited with the invention—or, more likely, the perfection—of Camembert around 1791. It was the introduction of the round Camembert box by a M. Ridel in 1890, however, that first allowed the cheese to travel further than Paris.

◀ A freshly-made Roquefort is very white, and, being only lightly pressed, rather moist—an ideal breeding base for the *penicillium roqueforti* spores. These new Roqueforts are in the Caves of Combalou—the only place in which true Roquefort may mature and blue. For as well as *roqueforti* spores in the air, the natural air-conditioning of the caves maintains a perfect temperature and humidity all year round. Of course, to ensure a full yield, spores are now artifically introduced into the cheeses; but nothing can replace the caves.

▶ M. Androuet, author of *Guide du Fromage*, puts his knowledge to practical use in Paris. As well as a cheese shop he has a restaurant which is renowned for its selection of cheeses. This is a typical display.

The white king

Brie is probably one of the world's most remarkable and romantic cheeses. Although to an Englishman Stilton is the king of cheeses, the rest of the world awards that title to Brie. And the accolade has been official since 1814.

During the Congress of Vienna, 1814-15, when post-Napoleonic Europe was being divided, much dinner-table argument was devoted to the question of which country produced the best cheese. At a grand banquet, the gourmet politician Talleyrand presented a perfect Brie de Meaux and it was unanimously crowned sovereign over the sixty other cheeses there.

The first recorded reference to Brie was at the court of Champagne in 1217. Louis XVI, foiled in his flight to freedom, with great aplomb, simply expressed a wish for some Brie and a glass of red wine. Dumas' three musketeers, exhausted by another day of chivalry and romance, often ended meals with Brie and Burgundy.

There are three types of cheese at which the French excel, and which are made to a much lesser extent by other countries: washed rind, goat and macerated cheeses. The washed rind cheeses are the ones with the bad reputations—the smelly cheeses. The process is explained on page 26 and certainly does produce cheeses with a most definite presence. But do not let this fool you. Most are sweet-tasting and are more mellow than the nose would suggest. The taste may need to be acquired, but is often an accurate taste picture of the pasture and flowers from which the cheese was made.

Goat cheeses are legion, from the tiny Chabichous to big, snowy logs. Their popularity is based, not just on flavour, but on the fact that goat's milk is easier to digest than that of cows. These cheeses can be

▶ French cheeses are made in every conceivable shape and size. The resultant flavours ensure that the Loire wines, the Rheims champagnes, the Bordeaux clarets, Provence reds and rosés, and fruity burgundies all find friends in local cheeses.

In the North the lush pastures of Normandy and the Ile de France give us the small flat Camemberts and white wheels of Brie, cylinders of herb-flavoured Boursin and tiny, rich Petit Suisse, heart-shaped Neufchâtel and round, red-rinded Maroilles.

In Eastern Haute-Savoie, Alsace-Lorraine and Franche Comté are the great Emmentals and Gruyères, Cantal and smaller St Nectaires. Western Brittany gives us Port Salut and St Paulin. In the Auvergne you will find mainly blue and flavoured cheeses. All over France, Chèvres are made in cylinders, logs, wedges—even pyramids.

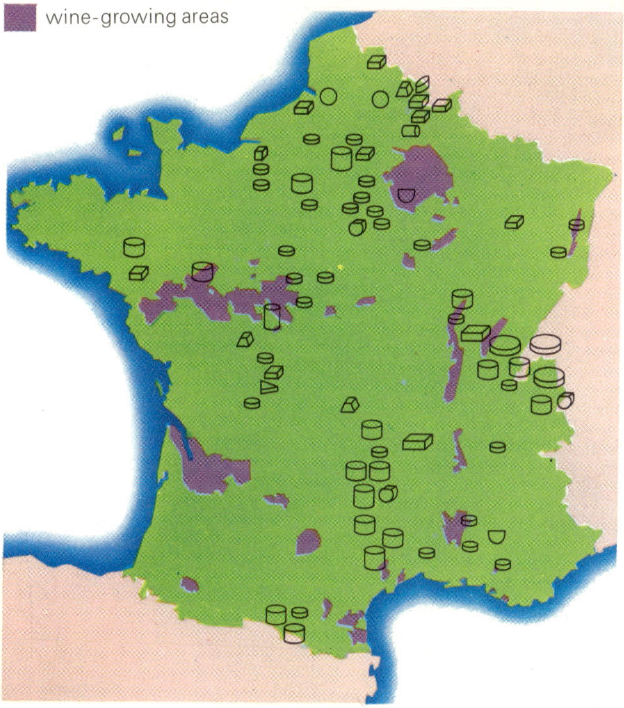

eaten fresh or aged almost to liquefaction. Many are covered in ashes to prevent the growth of mould. They are always fairly expensive, partly because the yield of goats is not high, and partly because the goat dairy farmer refuses to be organized into factory production.

Sometimes combination cheeses are made by adding cow's cream to goat's milk. These are the creamier *mi-chèvres*, often found flavoured with such herbs as summer savory and thyme.

Macerated cheeses are not seen abroad as often. Here the divine combination of wine and cheese is taken to its limits. Fresh—or sometimes mature—cheeses are aged or given their rind in alcohol—more often than not in *marc*, the fiery *eau de vie* distilled from the left-overs of wine-making.

Always individual, France has her own Gruyère and Emmental, rarely exported, and a Cheddar type called Cantal. There are bland cheeses produced by monks—the Port Salut of the Trappists—and of course Roquefort, made solely from the milk of the Larzac sheep. Now that sheep are mechanically milked, the basic white cheese is even made on Corsica and shipped to the caves of Combalou to be blued. It remains one of the world's great cheeses. It has been around for over 1,200 years—known by the Romans, enjoyed by Charlemagne and mentioned by the ubiquitous Casanova in his memoirs.

A catalogue of greats

There are perhaps about one hundred French cheeses exported in sufficient quantity to make it worth your while searching for them, and many are included in the following list. The majority are traditional, though up-dated. Every year sees the emergence of new cheeses, few of which offer something startlingly new or original. Therefore, a cheese you know and love may not appear here; this is not necessarily an oversight. It is more likely to be a rose by some other name. The comparative sweetness is for you to taste and then judge.

NOTES ON BUYING FRENCH CHEESES

Banon Laitier (Provence) Cow*: 45%: 2 weeks: 12m. Small disc, traditionally wrapped in chestnut leaves but now often in plastic. Lactic smell and refreshing sharpish taste when fresh; sometimes eaten when matured further. Banon in the UK should not smell high or ammoniacal. *Also made (less frequently) with goat and sheep milk, and/or sometimes marinated in *eau de vie*.

Beaufort (mountains of Savoy) PFCow: 5: 50%: 6m+. Best in winter, spring and summer. French Gruyère/Emmental type. Few or no holes. Fruity flavour and smell. Not often seen in UK.

Bleu d'Auvergne (Auvergne) P/R FCow: 9: 40-45%: 4-6 weeks+: P12. A cow's-milk Roquefort; sharp, salty and rich. Foil-wrapped cylinder; thin, pinkish skin. Avoid stickiness or over-bluing, which looks grey-green. Light and melting on the tongue.

Bleu de Bresse (Jura region) PFCow: 9: 50%: depends on size: P12. 3 sizes of small cylinder. Slightly mouldy smell, thin bluish rind. Is often creamy blue with a less sharp Gorgonzola-like piquancy. Avoid sharp smell, greyish body colour. Not one of the best blues.

Bleu des Causses (Aquitaine): very like Bleu d'Auvergne, but matured in caves. Another economical Roquefort substitute. Rather variable, though.

Bonbel A brand name for small, commercially made Saint-Paulin-style cheeses. The zip pack or wax covering means that it lasts well. Bland, mild, popular with children.

Bondon (Normandy) PFCow: 1c: 50%: 24 hours: P12. An unripened Neufchâtel (or Neufchâtel-type) cheese. Shaped like a small barrel bung. Fresh, slightly salty flavour, almost no smell. Avoid if smelly or soft. Its ancestry can be traced back 1,000 years or more in Normandy.

Boursault (Ile-de-France, Normandy) FPCow + cream: 2: 75%: 3-4 weeks!: P12. Also called Lucullus. Small cylinder, downy-pink rind, aromatic smell, mild, nutty, rich flavour. Avoid red surface, runniness or shrunkenness.

Boursin (Ile-de-France; Normandy) PFCow + cream: 1c: 70%: fresh: P12. Available plain, with garlic and parsley, or surrounded with black pepper. Rich, creamy and tangy. It is cheaper to make your own. Otherwise check the "sell-by" date on the pack, smell for rancidity. Avoid stained packs.

Brie Laitier (Ile-de-France, Burgundy, Champagne, Lorraine) PFCow: 2: 45%:

F/SPCow: 2: 50%: 3 weeks!: P12. The most common Brie, creamery-made. Comes in three sizes and two degrees of fat content. The white, floury rind may have some brown marking around the edge; smell should be that of mould, not ammonia. Texture smooth and homogeneous, with no chalkiness or excessive runniness. The flavour is full, mellow and almost buttery. It is incorrect (and dangerous) to eat ammoniacal or liquid Brie. Many of the packaged portions are excellent, but difficult to check on in supermarkets.

Brie Fermier (Ile-de-France) RFCow (occasionally pasteurized): 2: 50%: 4 weeks+. Summer, autumn, early winter. The most commonly available farmhouse Brie is Brie de Meaux. It is bigger than a Laitier with a much browner surface, even when immature. The smell and flavour are tangier, but still clean, lactic, with no trace of ammonia. Enjoyable even when slightly young. This was voted King of Cheeses in 1814 by the Congress of Vienna. Expect to pay quite a lot (up to 50% more) if and when you can find it.

Camembert Laitier (Normandy) FPCow: 2:45-55%: 2 months+: P12. The basic, creamery-made Camembert. Fat content varies with richness of milk. Small flat disc, white rind, some edge browning. Full mould aroma, not ammoniacal. Creamier colour than Brie. If perfect, should have no chalkiness or excessive coloration of rind, or liquidity Flavour varies from milky mellowness to a richer lactic. Not as strong as most people expect. Demi-Camembert and portions easier to check by feeling. Can be ripened at home in moderate temperature, but keep wrapped. Avoid high nose, shrunkenness in packaging or liquidity. Go for genuine Normandy product at first, though imitations are quite good.

Camembert Fermier (Normandy) RFCow: 2: 45-50%: 4 weeks!: late spring, summer, autumn. The real thing! Bears the description *"non-pasteurisé"* on chipboard container. Increasingly available, often made with milk from unsprayed grass. Like Brie Fermier it is browner and tangier, delicious when slightly young. Expensive, but worth every penny.

Cantal (Auvergne) FPCow: 4b: 45%: 3-6 months+: late summer, autumn. This fragrant grey-rinded, yellow cheese is often likened to Cheddar. Snap it up if you ever see it. If it is labelled "Haute Montagne" it will be particularly special and taste even more of Alpine herbs and flowers.

Caprice des Dieux (Champagne) PFCow + cream: 2: 60%: P12. Really an enriched Brie-style cheese, small oval-loaf shaped. Choose for individuality rather than exciting flavour, observing same caveats as for Brie.

Carré de l'Est (Champagne, Lorraine) PFCow: 2: about 50%: 3 weeks: P12. A flattish square Camembert-type, but slightly softer and milder. Must have snow-white rind, so no discoloration, no dryness or runniness. Relatively expensive when compared to finer cousins.

Chabichou (Cabichou) (Poitou) RFGoat: 10: 45%: 2 weeks+: scarce in winter. These small, truncated goat's-cheese cones are made in several ways. They have a strong animal smell and a very full flavour. Some farm-made examples can still be found. They should be firm and white-rinded but never hard, grey or blackened.

Chèvre This is a generic name for a host of goat cheeses which are as varied as the range of cows'-milk cheeses. The soft *blancs* should have a sharp sweetness, the logs should be firm, a little crumbly, and downy. Very few goat cheeses should be bought when actually hard, or with rinds that are markedly discoloured. Those with very blackened outsides are the *cendrés* and have been cured in cinders to inhibit surface-mould growth. Some of the more well-known cheeses include Chevrotin, Chèvreton d'Ambert, and Sainte-Maure. All have fat contents of 45%. Chèvres are easily digestible and form a good introduction to more exciting cheeses.

Comté French Gruyère (Franche-Comté): RF/S Cow: 4b: 45%: 3-6 months: end summer, autumn, winter. Although a Gruyère-style cheese, this is not cooked, and has large holes. It tastes like a stronger, fruitier Emmental and is rarely seen outside France, where it is enormously popular.

Coulommiers Petit Brie (Ile-de-France) PFCow: 2: 45-50%: 4 weeks+: P12. Basically a smaller Brie, sometimes enriched slightly, but with a taste closer to Camembert. Usually sold on straw mats and, in France, eaten relatively unripe, at the first appearance of the surface mould. Insecure packaging often encourages unacceptable dryness. Perfect dinner-party fare. Fair amount of browning acceptable.

Sometimes available made with unpasteurized milk.

Fondu au raisin—see *Tomme*

Fontainebleau (Ile-de-France) PFCow + whipped cream: 1c: 60%: P12. A rich creamy cheese, fairly firm. Flavour not pronounced, therefore some prefer it older, when it is sourish. Avoid a brown or slimy-looking outside. Many cheeses of this kind are marketed. St André is particularly rich—75% fat—and has a fuller, more satisfying flavour.

Fougeru (Ile-de-France): this is a Coulommiers-type cheese, usually made with unpasteurized milk, cured in a fern leaf. Some types of Fougère cheese (fern cheese) are now so rare as to be almost extinct. Excellent farmy lactic flavour, even when very young. A memorable experience.

Fourme d'Ambert (Auvergne) PFCow: lightly pressed: 45%: 3 months+: summer, autumn. Tall, slim, cylindrical: sometimes called a French Stilton. Pinkish-grey rind, salty, full flavour with intriguing after-taste. Some bitterness, light musty smell. Popular at Giscard d'Estaing's banquets. Avoid sticky, cracked rind, overbluing or grey/brown appearance. Bitterness and smell should not be pronounced.

Fromage Blanc PSCow: 1c: 0%: Sold under many names, Jockey and Bon Blanc being best known. Refreshing and popular as a sort of yoghurt substitute because of its very low or non-existent fat content. "Fromage Blanc" is also used, in France, as a family name for all fresh cheeses, even when full-fat.

Gaperon d'Auvergne PSCow: low: pressed, surface ripened: available year round according to degree of curing preferred. This rather expensive flattened ball of garlic-flavoured low-fat cheese is often made with buttermilk. It should have a greyish-white skin and is even acceptable with technicolor beard several inches long; but this is the old-fashioned way of enjoying it. Mildish garlic-and-pepper flavour. It should be a little moist outside or you will lose a lot of the interior. A conversation piece for a difficult dinner-party.

Livarot (Normandy) PF/SCow: 3: 40-45%: 3 months+!: P12. About the size of a Camembert but thicker; traditionally banded with sedge but now appearing with useless, coloured raffia. It has a strong, briny smell, slightly ammoniacal ("like town drains running into the sea"). The flavour is spicy and pungent, more than most other similar cheeses. Flavour and smell are caused by unventilated curing rooms. Difficult to tell when the smell and flavour begin to indicate *real* putrescence—and a very expensive mistake to make.

Maroilles (Hainaut, Flanders) PFCow: 3: 45%: 3 months+!: P12. Small, square and with much "presence". The reddish rind has an assertive smell, the pale shiny homogeneous body is vigorous and strong without being sharp. Ammonia should not be detectable, nor should dryness or runniness. Variations include Le Dauphin, flavoured with herbs and pepper.

Mi-chèvre This is another generic name for goat's-milk cheese that is either mixed with cow's milk or enriched with cow's cream. It makes a fairly light, rich aromatic white cheese when fresh, but it goes off rapidly. Often flavoured with herbs such as summer savory, then called Sariette.

Monsieur, Fromage de (Normandy) PFCow + cream: 2: 60%: 6 weeks+: P12. A small, white-furred cylinder of cheese that is yet another variation on the Brie-type cheese. It is richer in flavour, with a stronger smell. Avoid shrunken or excessively red examples. Invented at the end of the last century.

Munster (Laitier/Alsace) PFCow: 3: 45-50%: 4-6 weeks (small), 2-3 weeks (larger): P12. A famed cheese with a heritage said to date from the settling of seventh-century Irish monks—see *Swiss Appenzell*. Round, flat, orange-red rind and rich yellow body. The typical washed-rind cheese smell is penetrating, the flavour equally strong. Locally eaten quite young. The German version differs mainly by having an umlaut over the "u" in Münster. Some farmhouse versions are made.

Neufchâtel (Normandy) PFCow: 2: 45%: 3-4 weeks!: P12. In the UK we see more of the heart-shaped Coeur de Neufchâtel than the other variations. Its coat should be snowy-white and downy, with perhaps a little red pigmentation. The flavour is slightly more salty but, again, Brie is the best platform for comparison. Should never be sharp, runny or brownish. Excellent eaten young—see *Bondon*. Neufchâtel is a protected name.

Petit Suisse (throughout France) PFCow enriched with cream: 1c: 60-75%: eaten fresh. These small cylinders of fresh cheese should have a slight saltiness and a touch of sour-sharpness to enhance the light creamy taste. Avoid those showing signs of yellowness. Good with all fruit, great with raspberries—yes, with extra cream too, if you must!

Poivre d'Auvergne A delicious, modern Tomme, with moderate flavouring of black pepper. Horrible black plastic/wax coating.

Pont l'Evêque (Normandy) P/RFCow: 3: 30%: 6 weeks: P12. The real Pont l'Evêque about 4 in (10 cm) square is expensive (usually) and not too common in the UK for such a famous cheese. It should have a smooth gold-yellow rind, sometimes with straw indentations, a tangy sweet flavour and a moderately strong smell. Many variations include the larger Pavé from which better shops will cut a portion: even bigger—12 in (30 cm) square or more—is the Tour Grise. These are equally expensive, but you can buy smaller pieces. Same caveats as for all washed-rind cheeses.

Port du Salut PFCow: pressed 3: 2 months: P12. The original Trappist cheese, first made by monks in the thirteenth century. Although a washed-rind cheese it is very mild, almost bland. Excellent keeper. Ignore a slight growth of surface mould: salt water will remove it. Sometimes the growth is caused by exposure to excessive heat, so check for wholesomeness. A commercial version, Port Salut, is more commonly available. A wedge cut from the whole cheese tastes better than a prepacked wedge. Saint Paulin, although made slightly differently, is almost indistinguishable, other than being a little cheaper.

Reblochon (Savoy) PFCow: pressed 3: 45-50%: 4-6 weeks: P12. A flattened disc weighing just over 1 lb (·4 kg). The skin is pinkish-grey and smooth, with a slight mouldy smell. The creamy, sweet but throaty flavour is not as pronounced as might be expected. A relatively expensive cheese; unpasteurized farmhouse versions are very expensive. Worth it to experience an ancient *goût verité,* for the milk is processed when still warm from the cow.

Roquefort (Causse plateau) Sheep: 9/10: 50%: 3 months in caves of Combalou: 12 months by incorporating Corsican-made cheese. One of the world's most famous cheeses. A clean, sharp smell and pronounced, sheepy flavour with rich, salty after-taste. The bluing should be even, the body white. Although many cheeses are made with Corsican sheep's milk, these are actually ripened in Combalou. Corsican sheep lactate at different times of the year, thus ensuring us a year-round supply of the cheese. You might find it harder to track down in early winter. The texture should be buttery, but if you ask for a small portion from a block expect some crumbling of the edge. The locals eat it with butter; others think this heresy. Known since Roman times and eaten with chilled ham by Casanova.

Saint Marcellin (Isère) FPCow: 2: 50%: 2 weeks: P12. The modern cow's-milk version, about 3 inches in diameter, is made like Brie and Camembert but will develop a blue-moulded thin rind. Fairly mild with a little lactic bite. Served at the Louvre as early as the mid-fifteenth century. Then it was made with goat's milk, or a mix. Sometimes still found made like this in Savoie.

Saint Nectaire (Auvergne) PFCow: 3+ pressed: 8 weeks—better when much older. Available most of the year. About 8 in (20 cm) in diameter. Has a dry violet-to-pink thin rind with a slightly mouldy smell. Firm texture, fruity flavour with slight bite; but not strong. Farmhouse version still made. Melts well on toast; sometimes stirred into soup.

Saint-Paulin see *Port du Salut.*

Tomme (Savoy) PF/SCow: pressed 3: 45%: 2 months: P12. Tomme is another family which in general are uncooked, pressed cheeses which thus have a velvety-smooth texture and relatively mild flavour—sadly similar to processed cheeses, which they can resemble. The most popular in the UK is that with a plasticized black mixture of dried grape-skins and pips—Tomme aux Raisins. Cheaper versions can have a low fat content through being made with part or all skim milk. As some develop a sour or bitterish taste it can be difficult to judge the state. Authentic Tommes are often macerated in *marc* or flavoured, such as the Poivre d'Auvergne now on the market.
This family is also called Fondu.

British cheeses

▲ This ancient cheese press was photographed before World War II in Scotland. It was probably hundreds of years old and required a hefty milkmaid to handle it!

France may have given the world its king of cheese; England gave it the most popular and most copied: Cheddar. "Cheddar" has even become a verb, describing how curds are treated in the manufacture of a certain type of cheese. It is made in most western cheese-producing countries. At one time it used only raw summer milk, often coloured with saffron or marigolds, sometimes enriched with butter. English Farmhouse Cheddar could—and still can—offer unbeatable taste delights. Today's conglomeration of prepacked, pasteurized unripened Cheddars from France, Ireland, New Zealand or the United Kingdom offers all manner of shapes, flavours and textures. Many think this a good thing; the imitations, competent though they be, cannot fail to highlight the original.

Other English cheeses are older, and Cheshire was always better known and more highly regarded on the Continent. Even relatively modern French guides to cheese mention only Cheshire (*le Chester*), Cheddar and Stilton. These are grudgingly admitted to the hallowed, chauvinistic halls of French epicurean opinion; but the effect is spoiled by reference to Blue Cheshire as Blue Cheddar and on insistence that Stilton be moistened with port. If God had meant Stilton to contain port . . .

Quite as interesting as what Europeans think of British cheese, is why the cheese itself differs from that across the Channel, and why eating habits vary too.

Invaders and conquerors

England is perfect cheese country. The climate is just right; the superb summer milk from its rich pasturelands makes fine cheese, an important source of winter protein. As British shores were conquered and invaded, new cheese-making techniques and types of cattle were constantly introduced, giving the country an enormous variety of widely contrasting cheeses.

There were all manner of shapes and sizes. Hard, soft, blued, greened, herb-flavoured and coloured, herb-curded, goat and, in particular, sheep. There were big ones, small ones, round, square and oblong ones. Some travelled farther than others; a few were so common that they picked up several names—York and Cambridge cheeses were identical.

It is not too difficult to buy close approximations of what our ancestors knew—but only in the pressed and blue types. The soft and cream cheeses that were once a basic part of dairy management and domestic life have all but disappeared from the cook's repertoire. Even at the turn of the century there were not many recipes for curd puddings or cheesecakes which did not begin with the instruction to "earn" a quart, chopin or mutchkin of new milk—meaning to turn it with rennet. No popping round to the supermarket then!

The same books even gave recipes for making your own rennet. *A New System of Domestic Cookery* (the first nineteenth-century cookbook, incidentally, to allow the possibility of foreign recipes) presents a wonderful rennet flavoured with the evocative ingredients suggested only in such books: hawthorn tops, rose leaves, knotted marjoram, sweetbriar . . . what cheeses that might make!

English soft cheeses were flavoured with essences and spices; sweet herbs were layered between slices of velvety curd. Cheese was drained on split rushes, pressed in bricks, matured in leaves of dock, nettle and chestnut, or between plates of pewter.

The continuing Industrial Revolution, the railways and this century's international holocausts put paid to all that.

Farmyard to factory

Before the railways there was no alternative to home—or at least locally—produced milk. The poor state of the roads made transportation of raw milk impossible. Although cities and towns were geographically close to the country, cow- and goat-keepers were common sights around market places and town squares. Hygiene in the country was hardly of the highest; that in the towns was, literally, breathtaking. The rapid and not entirely unexpected spread of cattle plague through the town

▶ Cheshire cheese is one of England's oldest—the Romans even took it to Rome. Hence all over the country you will find old pubs like this one in Castleton named after the cheese. In London's Fleet Street there is a very famous one which still sells toasted cheese the way the great Dr Johnson enjoyed it.

▼ Picturesque Chester was the central point for the distribution of Cheshire for many centuries. Commercial production has meant the disappearance of traditional cheese fairs here, and cheese-making is now centred in Whitchurch, Shropshire.

herds in 1865-67 luckily coincided with the development of the railways. Trains were used to speed fresh country milk to the towns for the very first time. Concurrently, the revolutionary fever of big business led to the establishment of the first dairy factory in Derby in 1870. Others followed. No longer was the production of milk based solely on local demand and the productivity of the hired help. Indeed, who better to make this new factory cheese than the milkmaids and dairy-hands? The country lads and lasses followed their milk to the better conditions and pay of the creameries.

Down on the farm

Conditions down on the farm steadily deteriorated. They had neither spare milk nor hands to make cheese, not even the fast, soft cheeses for their own consumption. The creameries were not interested in these cheeses either, for how would they keep them edible? So they were forgotten, presumed to have no place in the modern world.

Then came the two world wars. During the first, more "dairy-maid" traditions and wisdom were forgotten. Resident bacteria evacuated themselves from old dairying-rooms and equipment, never to be re-identified or to return. During the second, rationing actually forbade making some of the few traditional cheeses left. In any case, the establishment between the wars of the new National Milk and Dairy Boards meant that farmers could earn as much for good plain milk as for going to the trouble of making it into cheese. Thus Blue Vinny, Cottenham (considered better than Stilton) and countless others, great and small, became extinct.

Even ten years after the end of the last war people were wondering whether they would ever see Blue Wensleydale or Sage Derby or Double Gloucester or Blue Cheshire again.

Throughout the long decades which witnessed the erosion of their traditional cheeses, the English—ever more willing to complain about the price of food rather than its quality—let them slip further and further away. Perversely, they served garlic- and herb-flavoured cheese from across the Channel, wide-eyed at the cleverness of the French, forgetting that a few years before they would have made it themselves—albeit not with quite *that* much garlic.

New pleasures

Things have improved tremendously. The glorious revival of interest in the home-made, the farm-made, the natural, whole and traditional has lead to a resurgence of interest in, and consumption of, farmhouse-made, single-herd cheeses. Financial strictures have also turned the thoughts of many firmly towards cheese and it is regaining its old importance as "white meat".

The whole farmhouse cheese-making industry is growing rapidly and in the last two years we have seen the reintroduction of two farmhouse cheeses not sold commercially since before the last war: Caerphilly and Sage Lancashire. Both perfectly illustrate the difference between chalk and cheese; the latter is fearsomely difficult to find, but oh! the pleasure when you do. Blue Cheshire and Blue Wensleydale are well re-established and Blue Vinny has been sold. Blue Vinny, however, it almost certainly was not.

Over the last fifty years, memories of the savour of this cheese have increased in inverse proportion to the amount seen and actually tasted. If you have been offered a Blue Vinny you would certainly have noticed a peculiar flavour, for it was the only major English cheese made with skim milk to have survived into this century. It was not veined in a Stilton or Cheshire manner, but clouded or streaked by a bacterium still not properly identified. It was said to have been made only from milk from around Sherborne; to set the seal on the unlikelihood of your having enjoyed the real thing, bear in mind that it was a very hard cheese indeed. You have probably been given an adolescent Stilton or an infantile Blue Wensleydale.

There is a growing interest in the recreation of some of the vanished soft cheeses, and with modern refrigeration and transport at our service, there seems no reason for the venture's failure. Between the wars there was a company flavouring its cream cheeses with chutney, caviar, lobster and an astonishing range of other flavours, some of them "compost-grown". Cost, if nothing else, one hopes, will prevent the resurgence of such excess. But the pleasure of sitting down to a genuine English country cheese, ripened to perfection and flavoured with naturally grown herbs, is certainly worth a little of someone's exertion.

English eating habits

Wine and the way he drank it is largely responsible for the Englishman's habit of eating cheese at the *end* of a meal rather than between the main course and the sweet course. Some erroneous views of the stomach's functionings contributed too.

The English owned Bordeaux for 300 years, from the mid-twelfth century, and they drank huge amounts of its bounty—but after, not with, the gargantuan mid-afternoon main meals of the day. Only soft or cream cheeses were included in the menu, and then often as an ingredient much fussed about with saffron, elderflower, spinach, tansy, rosewater and the like. Only the peasants ate hard cheese as a basic food, and most of this was made from the skim milk thought too rugged for the palates of gentlemen.

These same gentlemen might, however, consume a little hard cheese just before repairing to bed—to seal the stomach: they may also have nibbled some during their mammoth drinking sessions.

Gradually, eating and social habits changed. The main meal of the day was taken later and later, wine was served with it and eventually the small, light supper of cheese followed straight on from it. So there cheese stayed, still as a final stopping up. And is this not exactly the way the English still regard cheese? As a filler and topper-up? It is a 500-year-old echo of the belief that cheese's fat sealed the stomach, keeping the food and wine—be there ever so much of it—in place, while you slept and it did whatever it had to do.

The French always had more regard for cheese as a basic part of their diet. They served it in the Middle Ages as they still do, to finish the red wine that accompanied the roasts, and to signal a change of wine for the sweet courses. Fashionable Victorian society adopted a distaste for ordinary cheese. If we remember that only one in every three was supposed edible before pasteurization, this may have been good sense rather than bad fashion.

Now cheese of all kinds finds itself on all kinds of tables, at snacks, banquets and picnics. Yet old habits die hard. The English persist in serving cheese at the end of a meal, relegating it to an anti-climactical position after the cherries jubilee or spotted dick, rather than presenting it as a minor sensation before a triumphal dessert. Port is still thrown into Stilton and this cheese is still served with a scoop. And not so very long ago when faced with the sight of a Frenchman enjoying food and wine together, a native wit observed that an Englishman should never be expected to enjoy two things at the same time. Such feigned fragility belongs firmly in the extinct world of Oscar Wilde's drawing-room comedies.

Scotland and the islands

In spite of Scotland's long alliance with France, it has neither a palate for, nor a strong tradition of, cheese. Dunlop is the best known, but, like Derby,

▼ British cheeses are colourful as well as tasty. Blue Cheshire was known to only a handful until commercial bluing began after the last war. It is now regarded as one of the world's great cheeses. On the other hand, Red Cheshire is one of England's oldest cheeses, a favourite of the Romans. Its special flavour results from the saltiness of local meadows.

The mottling of crushed sage which emphasizes the pressed curds in Sage Derby is also a relatively modern invention. Less than twenty years ago the sage was a thick wavy layer in the middle of the cheese—you could enjoy mild Derby and pithy sage in the same slice.

Cheddar probably rates as the most internationally popular cheese. Farmhouse and creamery products taste very different.

it is fated usually to play second or third fiddle to the ubiquitous Cheddar, which it closely resembles. Enthusiasts deny this.

The Orkney and other islands have started more aggressively to market their interesting small, coloured cheeses, but they are not sufficiently distinctive to make much mark. Also their size, from about 1 lb (·4 kg) and up, makes them too small to be cut at delicatessen counters and too big for casual purchase.

Northern Ireland and Eire
The policy of the governments since the beginning of this century has been to stimulate the dairy industry as a whole and cheese-making in particular. So Ireland makes other people's cheeses, and generally makes them very well. Together with most of the traditional English pressed cheeses, she makes Emmental, Gruyère and Gouda as well as Brie and Camembert. Some old books, such as *Cheddar Gorge* (see *Books*), give recipes for cheeses based on the old traditions; these are worth making, especially if you live on the Emerald Isle, for those innumerable ballads about green hills are an accurate indication of the richness of the pastures.

▲ A Farmhouse Lancashire—still made today and usually eaten close to its place of origin. It is unique among English cheeses for using a mixture of curds from two different days. After three months' ageing it can actually be spread and has a considerably pronounced nose and flavour. In cooking it comes into its own, melting easily; in fact it was once called Leigh Toaster and is still best for making so-called Welsh Rarebit. A favourite companion in Lancashire is rich apple pie.

NOTES ON BUYING ENGLISH CHEESES

Caerphilly PFCow: 4a: 45-48%: 2-4 weeks: P12. White, granular texture, slightly crumbly; clean, sharp smell; sourish, buttermilk flavour. White/grey mould acceptable on farmhouse cheeses. Suitable ingredient, but an acquired taste, in cooking. Creamery Caerphilly is more sour and turns quickly; excess moisture or yellowing is bad.

Cambridge R/PFCow: 1c: 24-36 hours. Still occasionally made and available commercially. Small oblong with creamy top, orange central stripe, sitting on a straw mat. Fresh, slightly sharp but full flavour. Must be eaten very fresh. Also called York cheese.

Cheddar (Farmhouse) FPCow: 4b: 45-48%: 6m+: P12. Smooth, yellowish, waxy texture: full, sweet nose, rich nutty flavour. Slight after-taste bite which increases with greater maturity—as does crumbliness of texture. Useful in all cooking. Creamery or foreign Cheddars often have bite but no flavour owing to use of unripened milk or careless ageing. *NB* True farmhouse Cheddar is *never* made in oblong blocks.

Cheshire—Red FPCow: 4a: 45-48%: 6-8 wks+: P12. Crumbly, moister and less compact than Cheddar. Salmony pink. Mellow/sweet flavour with salty overtones; light clean smell.

White Faster ripening variation, uncoloured. Acidifies easily. Not much to recommend it.

Blue One of the world's greatest cheeses. Look for opulent marigold colour and buttery texture. Brown discolouring acceptable when close to rind. Quite unique flavour, very rich, full, nutty and tangy. You may like to temper it with butter. Surprisingly, many women prefer its robust flavour to that of Stilton.

Cotswold Double Gloucester cheese flavoured with chives; a fantastic modern commercial success. Rather artificial, moist, plastic texture. Good keeper. Like garlic, it is better if everyone at the table eats it.

Derby FPCow: 4b: 45-48%: 4 weeks+: P12. Flakier and moister than Cheddar, paler and more delicate in flavour, too. "Lesser Cheddar" is an apt but sadly damning description of this subtle and overlooked cheese. The first English cheese to be made in a creamery, in 1870.

Sage Marbled, flavoured and coloured with sage—and a few other additives. Highly individual, rather perfumed flavour; best appreciated in very small amounts. Interesting lightly grated over tomato salad.

Double Gloucester PFCow: 4b: 45-48%: P12. Yellow rather than pink, smooth mellow flavour even when mature. Must never bite. Particular affinity with fruit and salad. Single Gloucester—also called Berkeley—was a faster maturing cheese of the same diameter but half the depth. Sadly extinct.

Ilchester Another modern conglomerate. Tremendously popular in the 'sixties and now finding favour again. A mixture of Cheddar, beer, chives and garlic.

Lancashire PFCow: 4b: 45-48%: 2m+: P12. White, crumbly but butter-soft texture. A unique blend of curds from several days. Mild but rich flavour and *the* cheese for cooking and melting. Mature Lancashire has a wonderfully full flavour.

Leicester PFCow: 4b: 45-48%: 6m+: P12. Rich russet-red colour, granular looking but actually moist, elastic texture. Clean buttery flavour. Often marketed too young, as the flavour develops before the texture. So avoid dry, crumbly-looking cheeses.

Stilton—Blue PFCow: 9: 45-48%: 6m+!: P12. Soft buttery texture, blue-green veins radiating from centre to crust. Wrinkled brown/grey skin. Some brown discoloration close to the skin is acceptable but nowhere else; it is a sign of inferior milk or bad manufacturing. Neither flavour nor smell should be sharp or strong. Do not accept white and chalky (unfortunately not a rare commodity) or yellow and oily Stilton. One is too young, the other too old. Can, but should not, be used in cooking. Port is an accompaniment, not an ingredient.

White Crumbly, very white cheese with strong nose but deceptively milder, slightly sour, flavour. Ideal substitute for Fetta in Greek salads—but that's about all.

Walton Astonishingly popular mash of Cheddar, Stilton and walnuts. The murky look and musty flavour are intentional. Refined manufacturing will improve the appearance.

Wensleydale—Blue PFCow: 9: 45-48%: 6m+!: P12. Though smoother, whiter, smaller and less veined, it is undeniably similar to Stilton. Grey corrugated crust. Wensleydale curd is never soured, giving it potential for a richer, rounder flavour. A perfect example could well rival Stilton.

White A lightly-pressed, rapid-maturing cheese that has a definite character of its own. But appreciation of its subtle, mellow, honey-like flavour requires the possession of a fairly well-tuned palate. Sourness and yellowness indicate excess age. It has no rival as an accompaniment to apple pie.

Windsor—Red A Cheddar-based cheese with hysterical marbling of electric-red liquid, reputed to be wine. A fit mate for Walton (qv).

Scotland

Caboc PFCow: 1c + oatmeal: ?: fresh: P12. Newish, small full-cream cheese coated in oatmeal. Slightly sharp nutty taste; check carefully for mould or deterioration on the oatmeal.

Dunlop PFCow: 4b: 45-48%: 2-3 months+: P12. Like Derby, this popular Scottish cheese is similar to Cheddar and to Double Gloucester. It is paler, blander and moister. Naturally enough it is best with Scottish accompaniments, oatcakes in particular.

Orkney and Islay Small soft cheeses with an authentic provenance but such a similarity to better-known English cheeses that their purchase is on impulse rather than cultivated demand. Delicious, nonetheless.

Greek and Italian cheeses

GREECE

Ancient Greeks regarded knowledge of cheese-making as a gift from the gods and honoured the lonely life of the vital shepherds and shepherdesses in song and story. The priapic gods and their amorous consorts paid them a good deal of attention too, offering a kind of job satisfaction that was nothing if not heavenly. True or not, cheese has certainly been part of the Greek diet since history began. Homer recounts that the one-eyed giant Polyphemus, the captor of Ulysses, had his cave stacked with rack upon rack of cheeses and barrels overflowing with whey. In Athens cheese was a luxurious and rich commodity; in the sweaty gymnasiums and open fields of athletic Sparta it was a simple essential in a libido-controlled diet.

The poverty of much of Greece's soil is such that there was never much hope of raising animals for both meat and milk: it was one or the other. Thus the Greek diet has always relied on fruit, cultivated grain, olives, wine and cheese. Meat-eating was for important occasions, and still is. Today the Greeks eat more cheese per head than any other race.

Greek cheese is limited in range and simple in manufacture. But its individuality—achieved by only the slightest of margins—has led to a tremendous upsurge of interest and consumption in the past few years. European demands, mainly from nostalgic holidaymakers but also from displaced nationals, have caused a dairying phenomenon—Denmark has begun manufacturing and exporting Fetta. Of course, it is not quite the real thing, but then neither is a great deal of what you may buy in Greece nowadays.

Fetta is the most universally known Greek cheese. Originally and authentically made only from sheep's milk, it is the result of quite a fascinating process. The milk is allowed to coagulate naturally (acid-curding). Then the curds and whey are heated, which causes flocculation of albumen and any other remaining protein.

Most modern Fetta is an approximation of the real thing, and almost certainly made from cow's milk. To be acceptable to the Greeks, this must be bleached to give the finished cheese an acceptable whiteness. But as long as the lightly pressed curd is allowed to ripen in its own whey, diluted with brine, it will develop an acceptable flavour.

Mitzithra is becoming quite well known, too, most often in the Greek kebab houses which have sprung up throughout London. It is a whey cheese, similar to Ricotta, but is now probably enriched by the addition of milk. Throughout Greece and her myriad islands, goat's milk and sheep's milk are used separately or mixed together. The resultant unexported and sometimes unnamed cheeses are used not only as an essential part of almost every meal but as the starting point for some spectacular but simple recipes.

NOTES ON BUYING GREEK CHEESES

Fetta RFSheep*: 1b+ pickling: 45-60% fresh or ripened: P12. Genuine sheep's-milk Fetta is made from naturally curdled milk which is then heated to flocculate the remaining casein and albumen protein. It is then cured in a mixture of its own whey and brine.

Commercially-made Fetta is usually made with pasteurized cow's milk, sometimes bleached, sometimes with sheep's or goat's milk added. In any case it should always be very white, moist and slightly crumbly. Taste is salty-sour, but refreshing. Avoid excess generation of moisture or separating; never buy the pre-packed cheese if the plastic pack has swollen.

Most Fetta seen in England is from Denmark, a consequence of the difficulty in importing the authentic Greek version in good condition.

Mitzithra RFSheep*: 9: low fat content: fresh or cured. This is a Greek version of Ricotta and, like the Italian whey cheese, is subject to infinite variation, since full or skimmed milk may be added to the hot whey to increase both the yield and the richness of the product. May be dried so that it is suitable for grating. Commercial Mitzithra is usually made from the whey of pasteurized cow's milk.

ITALY

Cheese was a major product well before the great Roman Empire. Sheep and goat herds were the most popular, as they required less attention and feed than cattle. Hygiene was acknowledged and practised. *Agricultura*, the oldest known Latin book, continually stresses its importance.

Cheese was an important part of a legionary's pack while marching across Europe. It is thought that these same legionaries were largely responsible for introducing more sophisticated cheese-making to Britain.

When the Empire faded, cheese-making followed suit. Nothing much more than farmhouse soft cheese was made. But the knowledge spread first by soldiers and later by the proselytizing monks of the new Christian religion fostered Rome's cheeses in foreign lands. Indeed, today it is quite possible to trace some cheeses' ancestry directly back through these routes. It was well into the 10th century AD before there was a revival of more adventurous cheese-making in Italy.

Bullet-proof and plastic

Parmesan (*Parmigiano*), one of the Grana family, is probably the most authentic pre-Roman-style cheese still made. Very dry and salty, it is relatively low in fat, being made with skimmed milk. Some take many years to mature and it is claimed that the oldest and ripest Parmesan is actually bullet-proof.

The blue cheeses of Italy are all based on Gorgonzola, one of the earliest products of the renaissance of Italian cheese, and known for over a thousand years. Once the butt of jokes because of its smell, it is now a mellower item altogether, mainly owing to the use of pasteurized milk and a shorter maturation period. It has two claims to being unique—it was always made with pure cow's milk and the blue mould, although of the penicillin family, is not of the Roquefort strain.

Pasta Filata is the Italian generic name for a weird and wonderful range of plastic curd cheese—Provolone, Caciotta, Caciocavallo, and others. These strangely shaped straw-yellow cheeses usually hang in pairs like equipment for esoteric team games. Those exported or made abroad are universally soft, rather bland but sweet cheeses.

Production of goat's cheese has declined drastically, other than of delicious Caprino, a soft cheese. But sheep's-milk cheese continues, mainly in the form of Pecorino, sold both soft for eating and hard for grating.

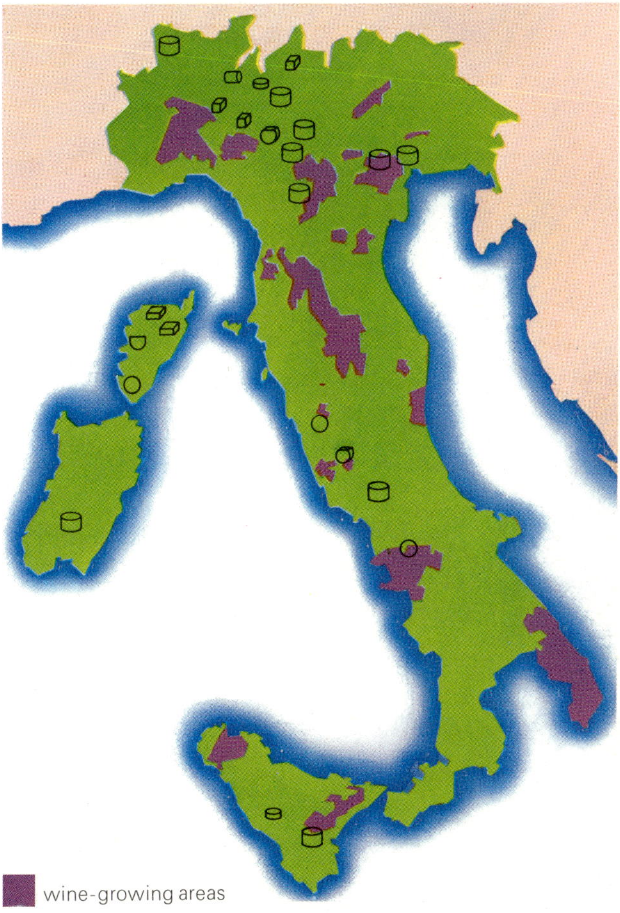

▶ The excellent climate and rich soil of northern and central Italy are largely responsible for the long establishment of sophisticated wine-making and a gourmet cheese range. Salty Parmesan and Grana or tangy Gorgonzola perfectly partner red Chianti and Valpolicella; blander Provolones and rich cream cheese mate with such famous whites as Soave and Frascati.

In the south, a harsher climate and more rugged countryside inhibit wine- and cheese-making by producing lower yields and making long-term storage almost impossible. The heady local wines are drunk young; cheese is usually eaten fresh, only sometimes being pressed and stored until suitable for grating.

NOTES ON BUYING ITALIAN CHEESES

Bel Paese PFCow: 3: 28-29%: 6-12 weeks: P12. Firm pearly-white texture with full fruity nose. Clean creamy taste, sweetish after-taste. A mild, washed-rind cheese perfected about 1920 in Lombardy. Great keeper; but beware of an excess of paraffin under the foil, a fault easily detected by your nose.

Crema Bel Paese is a very soft, presumably processed, version of Bel Paese, sold in small round packs.

Dolcelatte see *Gorgonzola*: this is a milder, creamier version of that robust favourite. In essence, a providential combination of modern factory production and marketing techniques, but no less good for all that.

Fiore Sardo see *Pecorino*.

Fontal is Fontina (see below) made with pasteurized milk.

Fontina RFCow*: 5: 45%: 3 months+: July to December/ January. The great Piedmontese cheese, slightly yellow, sweet aroma, delicate nutty flavour. A few small round eyes. Melts well and grates when matured and hard. Once made with ewe's milk, and sometimes still is.

Gorgonzola PFCow: 9: 48%: 90 days+: P12. Soft texture and blue-green veins in a yellowish paste. More elastic than other blue cheeses. Sharp smell and rich but not pungent taste. Beware a brown appearance, bitterness or sourness. Made for over a thousand years.

Grana Basically the same as Parmesan (which is, in fact, a Grana) in manufacture, but both names are strictly controlled according to area of manufacture. Whereas Parmesan Parmigiano is made north of the Po, Grana is a family of similarly produced cheeses made all over Lombardy. Grana is generally lighter in colour, faster to mature and flakier. Excellent in cooking as it does not form threads. Grana Padano, made with pasteurized milk all year round, is widely distributed.

Mozzarella PFCow*: 6: 50%. Pale and plastic-looking; unobtrusive smell and flavour. Eaten raw (try it sliced and marinated in vinaigrette, with orange slices and a few chillis, dry or fresh) or cooked on pizza, etc. Should be kept in light brine if paper-wrapped, refrigerated if plastic-packed. Originally a buffalo-milk cheese.

Parmesan (Parmigiano) RS/FCow: 5: 32%: 2-3 years+. The salt and pepper of Italian cuisine, now prohibitively expensive. Although hard and crumbly, Parmesan literally melts in the mouth with a rich, full, pithy flavour that belies its very low fat content. Young Parmesan, *Giovane*, is a truly delicious table cheese; *Tipico, Stravecchio*, and *Vecchio* are progressively older, harder and more highly flavoured. More costly too. Ready-grated Parmesan *must* be lesser flavoured and, if you work it out, breathtakingly expensive. You would be better off using Pecorino or mature Cheddar.

Pecorino RF Sheep: 4a: 35-40%: 2 months+. Pecorino is the family name for Italian sheep-milk cheeses, the most famous of which, Pecorino Romano, is rarely seen in the UK. But the Sardinian variety, Pecorino Sardo *(Fiore Sardo)*, is quite easy to find. It is a golden yellow wheel with a convex rim about eight inches in diameter, with rush or straw markings, sometimes artificially added these days. Similar nutty flavour to Parmesan and a good substitute. Excellent table cheese and keeper. Grates well even if fairly soft; store wrapped in moist cloth in refrigerator.

Provolone PFCow: 6: 44-47%: 6-9 months. Seen in several sizes and shapes but usually a small (6 in/15 cm) or large (18 in/45 cm) truncated cone or egg-shaped loaf, up to 6 in (15 cm) across. Glossy golden rind, dense creamy paste, delicate and mild when young, piquant when old. Must not have internal holes.

Ricotta Cow/Sheep/Goat: 7: 20-30%: Fresh. Once made simply by heating whey, but nowadays full or skimmed milk is usually added for a greater and richer yield. It is fresh, white and crumbly, slightly sweet and similar to cottage cheese. Can be sweetened, but usually is mixed with other ingredients to stuff pasta. It may be dried and grated as a condiment. Sours rather easily.

Taleggio PFCow: 3: 48%: 6-8 weeks: P12. This was created either early this century or a thousand years ago, depending on the book you have in your hand. It is a creamy, full cheese with a subtle bite. The rind should be a pinkish-grey, but is often seen in the UK with rather a lot of blue-grey mould. Provided the body of the cheese is relatively firm and evenly coloured it is still an acceptable purchase, but it should not be misshapen.

German, Dutch and Swiss cheeses

The cheeses of Germany, the Netherlands and Switzerland have more in common than a first glance and their incredibly contrasted scenery might suggest.

Common borders, common traditions

With Germany and the Netherlands on the route of every would-be conqueror, and Switzerland the traditional neutral refuge of those wanting to avoid such conflict, there has long been an exchange of cheese lore. The early establishment of international trading centres in Holland and Germany also guaranteed constant exposure to advances in techniques.

It is impossible to make the same sort of cheese with the milk of the great, brown Swiss cows as with the pied Friesians of the lowlands. The grass is different, and life in the Alps—where dairymen and their cows disappear to the high slopes for the entire summer—has nothing in common with that in countries where the herd is usually on your back doorstep and often actually under your roof. Cross-fertilization of ideas and techniques rather than blatant copying led to the very early consolidation of all of these countries' basic cheese styles.

Switzerland has been an exporter of fine, long-lived, hard-pressed, cooked cheeses since the thirteenth century, when the St Gotthard Pass opened up a trade route with Italy and the rest of Europe. Cheeses were much smaller then, but advances by the mid-sixteenth century saw great golden cartwheels of cheese floated by barge down the Rhine and to the maritime markets of Holland.

▲ Swiss farmers gather to share cheese communally made and stored.

▼ Edam and Gouda are auctioned by medieval guilds at Holland's Alkmaar Cheese Market.

These cheeses were traditionally made high in the mountains by only a few men, the Sennen. Most of the dairymen were fully occupied in watching the cows, so the yield from each milking was pooled and made into cheese at a few central points. The cheeses were stored communally until it was time to lead the cows back to the valley and safety during winter. Two very important celebrations are associated with this time of the year, both of which are often conducted in the traditional manner. One is the great sharing-out of the cheeses. The other is the formal procession of the milking animals down from the Alps—the *Alpabfahrt*. These long, meandering lines of belled cows, some decorated with truly fantastic headdresses, are welcomed by the whole village—joy no doubt associated in no small way with the return of cowherd husband to wife.

Holes and herbs

There is much confusion about which Swiss cheese has the holes. If it *is* Swiss, it's Emmental. If it is French, then Gruyère has the holes and Emmental is blind. There is also an important difference in the cooking attributes of Gruyère and Emmental. Blander Emmental tends to draw threads when heated; fruitier Gruyère does not. A mixture of the two produces a Gouda fondu with a well-balanced flavour and texture.

Nowadays such cheeses are no longer an exclusive product of the mountains, for it has been discovered that valley grazing, if beefed out with modern foodstuffs, produces milk and cheese indistinguishable from the "real thing"

Mountain-made Appenzell has a peculiar claim to fame. It was introduced to Switzerland by eighth-century Irish monks, and St Brendan is credited with the first European discovery of the American continent at the same time. The 1977 expedition by the Irishman Tim Severin to resail the route in an open leather boat included the bi-national Appenzell in its stores, to counteract the monotony of a diet of smoked meat and salted fish.

Holland's famous Goudas and Edams were amongst the very earliest international bestsellers. Organized dairy-trade centres were set up in mid-thirteenth-century Haarlem for the sale and distribution of cheese. At the same time, British merchants were regularly buying cheese in Gouda.

Although there is a town called Edam, today it is Alkmaar upon which Edam cheese trade centres. While it is one of Holland's most important and colourful tourist attractions, it still serves a very necessary and commercial purpose. Each Friday you can see straw-hatted cheese-bearers offering their shining wares for sale from long wooden sledges, just as in medieval days. Most other authentic Dutch cheeses are simply variations on the Gouda and Edam themes.

Germany's most popular cheeses—at least in terms of national consumption—are fresh cheeses of varying fat content, all strictly controlled, and Quark, a low-fat curd cheese. Quark is used in every facet of cooking and finds its way into every type of comestible combination.

Far better known abroad are three smelly cheese styles, one of which is genuinely German, the other two of which are not. The first are the hand cheeses—Handkäses—made from soured Quark, or Quark made with sour milk, or from whey treated in the same manner as Ricotta (see *Travellers' cheeses*).

The equally smelly Limburger and Münster cheeses were originally Belgian and French, respectively, but the Germans have made such a success of their manufacture, particularly of Limburger and its derivatives, that they are now regarded as authentic German cheeses. German Münster has an umlaut; otherwise it is French Munster from Alsace-Lorraine.

▶ The return of the herd, garlanded with paper and flowers at the end of the summer.

NOTES ON BUYING DUTCH, GERMAN AND SWISS CHEESES

Edam (Netherlands) PF+SCow: 4a/b: 40%: 4 weeks+: P12. Whey and brine baths and salting complete the rind of this internationally famous cheese after it has been pressed at medium pressure. It is in fact full-flavoured despite its low fat content, but is usually sold well before this flavour has had time to develop. Has been called the cheese for people who hate cheese, owing to its sometimes innocuous flavour. Sourness is a give-away sign of youth. Only export cheeses are coloured red. Quite useful in cooking; good filler, too, as it is still one of the cheapest cheeses around. Remarkable keeping properties.

Gouda (Netherlands) PFCow: 4a/b: 48%: 2-3 months+: P12. Richer, larger, yellower than Edam, with a bigger flavour, too. Salt rubbed on the rind helps develop the flavour, which continues to improve for over a year. Mature Gouda, usually coloured gold, is a real treat and tastes like a completely different cheese. Excellent cooker whatever the age, and a reasonable substitute for Gruyère and Cheddar. Melted with milk it makes a Dutch fondue— the *Kaasdoop* or cheesedip.

Leiden (Netherlands) PF+S+buttermilk: 4a: 30-40%: 2-3 months+: P12. This cheese is kneaded when in the curd stage (in former times it was trodden). Some farmhouse cheeses still made. Usually flavoured with cumin.

Limburger (Germany, but originally Belgium) PFCow: 3: 30-50%: 13 weeks!: P12. Oblong, loaf-shaped; bigger than most washed-rind cheeses, therefore there is more of the brick-red rind to assail your nostrils. Usually foil-wrapped. The rind should be smooth and only just moist; the yellow, smooth-textured body has a very full flavour, less assertive than the farmhouse odour might suggest. Beware ammonia in the penetrating smell, shrunkenness and firmness.

Mainzerkäse (German Hand-käse—Sauermilchkäse) PSCow: 1b pressed: varying: P12. These opaque yellow cheeses are made from naturally curdled sour milk, and hand-shaped into various forms. Those usually found in the UK are small, indented logs wrapped in cellophane. They have a penetrating smell and the close-textured body is, at the very least, piquant and usually downright over-powering.

Münster (Germany) PFCow: 3: 45%: 5 weeks! or 13 weeks!: P12. Made in several sizes, hence the difference in curing times above, but all are flattish discs. White-to-yellow body, thin skin rather than rind, strong smell but slightly sharp sweetness of flavour. Watch for dryness, cracking, etc., or very moist slimy skin.

Appenzell (Switzerland) PFCow; 5 with rind treatment: 45%: 3 months+: P12. This Swiss cheese has no holes, is rather firm, and can be as mild as an Emmental or fruity as a Gruyère. During curing it is regularly washed in brine flavoured with wine, cider, herbs and pepper. It is good in cooking, usually as an addition. A well-ripened Appenzell has a rich sharpness which may be better appreciated in a hot dish rather than on its own. It should not be cracked, hard or dry. Fairly rare in the UK; usually found pre-packed rather than whole, when it is about 14 in (35 cm) across and only a few inches thick.

Emmental (Switzerland) RFCow: 5: 45%: 6-12 months +. Additional use of pasteurized valley milk means year-round manufacture and supply. These great wheels can weigh almost 200 lb (80 kg), in spite of all those great holes or eyes. The eyes should be fairly evenly distributed, the body a light yellow, the rind smooth and free from mould (which may, nonetheless, easily be wiped off, and causes no ill effect). The smell and taste should be sweet and clean, not bitter or assertive in any way.

Useful in cooking but will draw threads if heated too highly. Real Emmental always has "Switzerland" stamped on the rind in red. A drop of moisture or "tear" in the eyes is a very good sign of a perfectly mature and well-developed cheese.

Gruyère (Switzerland): PFCow: 5 with washed rind growth: 6-12 months+: year-round manufacture and availability. Only half the size of Emmental but different curing and the development of rind flora gives a much fuller, fruitier flavour. The texture is supple with very few holes, no larger than a cherry. Check for the "Switzerland" stamp on the surface. Some sliminess on the rind can be dealt with and ignored, as long as the cheese has not become bitter or sharp. Beware of gritty or grainy texture, sometimes brought on by cracks on the rind. Great cooker; hardly draws threads at all.

Scan-dinavian cheeses

Most of the food consumed by the Vikings was salted and smoked to a uniform degree of noisomeness—except their cheese.

A separate tradition

Northerly pastures boast little of the sweet-tasting grasses or savoury herbs of the warmer south and so produce less rich milk. This was a blessing to those old Scandinavians, for they could preserve it just as it was, with minimal salting and no smoking. Imagine the relief of biting into soft, sweet cheese after a heavy meal of salted raw fish and smoked mutton.

Such bland cheeses have long been the mainstay of native production and diet. Then, of course, there is whey cheese; quite another story altogether.

An important product of any thrifty, Scandinavian dairying community, this golden-brown milk by-product is not strictly a cheese and has a pungency and individuality of flavour that allows no indecision—you either love it or hate it. Called Mysost in Norway and Mesost in Sweden, it is made by boiling down whey from cow's milk. But rather than stopping when the solids flocculate, as with Ricotta, the whole mass is kept on the simmer until all extraneous whey liquid has evaporated. The brown colour and peculiar bite of these cheeses results from the caramelization of the lactose. If all or some of the milk is from goat, the cheese becomes Gjetost, a Norwegian speciality and perhaps the best known of these cheeses. It totals 30 per cent of that country's cheese production.

A diet staple

The Swedes, Norwegians and Danes, as well as their neighbours to the south, the Dutch and Deutsch, eat a great deal of cheese. Even at breakfast time it is scraped paper thin with a special implement and served with salamis and cold meats. Nowadays their choice of cheese runs the full gamut of flavours, mainly because of wide experimentation in the nineteenth century and the influence of advisers from Switzerland and Holland.

Danablu, or Danish Blue, is probably the best example of a new Scandinavian cheese. It was invented as a purely commercial venture early this century and proved internationally successful. This story is typical of the forward-looking Danish dairy industry. Danish herds were the first in the world to be tuberculosis-free, and the current standards of hygiene are exemplary. As far as cheese of the future is concerned, Denmark is certainly a country to watch.

Norwegian and Swedish cheeses are not widely exported or known, with the exception of Norway's Jarlsberg. Once it was on the point of extinction, but modern scientific research rescued it; modern marketing has made it popular. It is to Norway we must also give our thanks for the idea of communal creameries and co-operatives. They set up the world's first in 1856.

▼ Cheese knives are almost unknown in Scandinavia. Instead, cheese is sliced very thinly with the special implement shown below. Here the cheese is Gjetost, the brown whey cheese that appears at most meals, including breakfast.

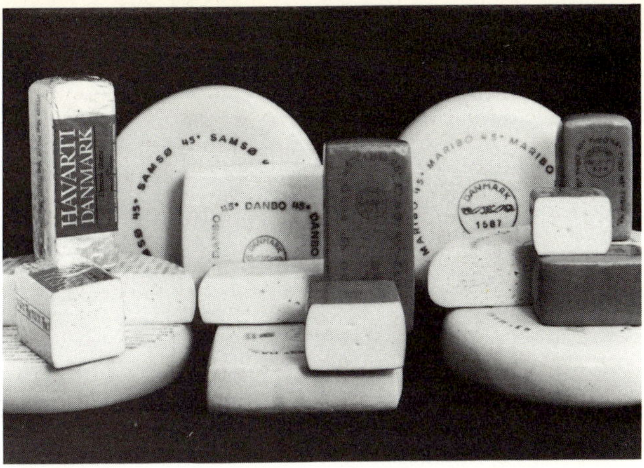

◄ Superb hygiene, excellent milk and willingness to learn from others has given Denmark a great range of national cheeses; she also makes those of other countries.

NOTES ON BUYING SCANDINAVIAN CHEESES

Castello: see *Mycella*

Danablu (Denmark) FPCow: 9: 50-60%: 8-10 weeks: P12. Invented in this century. Sharp saltiness belies its richness, due to the use of homogenized milk. Should be white rather than creamy, with blue rather than green veins. Crumbliness acceptable if there is plenty of moisture. Often used as a substitute for Roquefort in cheese mousses, blue-cheese dressings and other Americanisms. Mixed with butter it makes an excellent savoury spread.

Danbo: see *Samso*

Elbo: see *Samso*

Esrom (Denmark) PFCow: 3: 45-60%: 4 weeks: P12. A Danish Port Salut but with a much more aggressive mind and presence of its own. Firm and buttery with small holes. Rich, sweet flavour which continues to develop. Aficionados swear you *must* eat the rind.

Gjetost (Norway) Same details as Mysost but made with goat's-milk whey or cow's and goat's mixed. When pure goat's-milk whey is used, it is called *Ekte* (real) *Gjetost* (even more an acquired taste!). Sliced very thinly, this nutty-flavoured cheese is a great complement to herring and fish dishes.

Havarti (Denmark): FPC: 4a+3: 30-60%: 8 weeks: P12. There are two kinds of Havarti – that with a dry rind and that with a washed rind which develops a sharper flavour. The dry-rind version is similar to Tilsit and was once called Danish Tilsit. Aromatic but basically bland flavour. Small holes. Made in flat rounds or loaves.

Jarlsberg (Norway) FPCow: 45%: ?: P12. An old Emmental-type cheese revived this century and widely marketed. Large holes in creamy yellow body. Soft, smooth and mildly aromatic, but less so than Emmental.

Mycella (Denmark) FPCow: 9: 50%: 8-10 weeks!: P12. The Danish Gorgonzola; full but unaggressive flavour, yellowish, supple body and green, not blue, veins. Avoid brownness or over-bluing, sharp smell or bitter flavour. Castello is Mycella with a surface, Camembert-type mould as well, which is intentional. A great success in Australia.

Mysost (Norway) PFCow: 7: 10-33%: fresh: P12. This pungent, brown whey cheese is not truly a cheese at all. Addition of whole milk or cream adds to the fat content. Usually sliced thin but is used in Scandinavian cooking, particularly sauces. Can be stored quite well.

Samso (Denmark) FPCow: 5: 30-45%: 6m+: 10 weeks: P12. Originally a Danish Emmental substitute, but now recognized as a distinct family, probably Denmark's most important. All the cheeses with "o" on the end are related; the first part of the name indicates the cheese's origin. Samso is firm, yellow and nutty with cherry-sized holes and a flavour which varies from mild sweetness to distinctive strength. It is made in a cartwheel shape.

Preference is highly personal and rather arbitrary; varying fat content may have some bearing. Very useful in cooking and probably best defined as between Cheddars and Emmentals.

Danbo is square and sometimes caraway-flavoured.

Elbo is rather firm, loaf-shaped, and has few holes.

Tybo is brick-shaped, very firm and red-rinded.

Tybo (Denmark) see *Samso.*

Travellers' cheeses

Generally speaking, most countries offer national favourites not exported—or at least not in any great quantity. Like many fresh, simple country wines, these cheeses may not travel well—or perhaps it is just that fresh mountain breezes and a glass of the local vintage give them a character that no export market can reproduce. Certainly the adage "If it's good, it's shipped" is not necessarily true. Often these local cheeses are so good that there is none left over for foreigners.

Regional specialities

The French Gruyères, Comté Beaufort and Emmental are made in vast quantities but consumed almost exclusively in France. In Germany, besides the ubiquitous Quark and other *Frischkäses,* there are limitless varieties of *Handkäses*—discs, cylinders and squares—which change from village to village and farm to farm. The more "nosey" of these will actually drive grown—but uninitiated—men from the table.

The Scandinavian and North Sea countries do not have many rarities to offer, but you will find it worthwhile looking for Gammelost in Norway. This is a smallish cheese with an ancient history. During its manufacture the long furry surface mould is worked back into the cheese.

Holland, Belgium and Germany all offer Limburger variations, and the cumin-and-clove Nagelkaas of the first is a delight to be tried on a visit to Friesland.

A trip to Switzerland gives you the chance to enjoy Raclette. Raclette is not a type of cheese, but a generic name for a group of mountain-made cheeses which melt well. All skiers know the pleasure and companionship of a Raclette feast around an open fire after an exhilarating day on the slopes. Unfortunately there is no real substitute for Raclette cheeses – a fact sadly related by Swiss all over the world who have dutifully experimented with local cheeses. Schabzieger is "condiment" cheese which deserves attention. It is a very hard, greenish cone that is usually grated on to buttered bread. It is quite pungent and actually smells like meat-extract cubes. Used responsibly it is different enough to become addictive, and is thus exported, in small quantities, sometimes already grated.

Italy boasts a number of curiosities, many of them of the

◀ In most hot countries with predominantly peasant-worked land, cheese is usually eaten fresh. Temperature and living conditions make storage virtually impossible; limited sources of food make it difficult, in any case. But intrepid travellers, such as those visiting this Turkish family, will always find themselves eating freshly made goat- or sheep-milk cheese, offered with justifiable pride.

▶ "America's Leading Cheese Stores" are the Hickory Farms of Ohio, with branches throughout the United States. They have pioneered large-scale cheese retailing, and carry most of the cheeses—imitation and indigenous—produced throughout the USA.

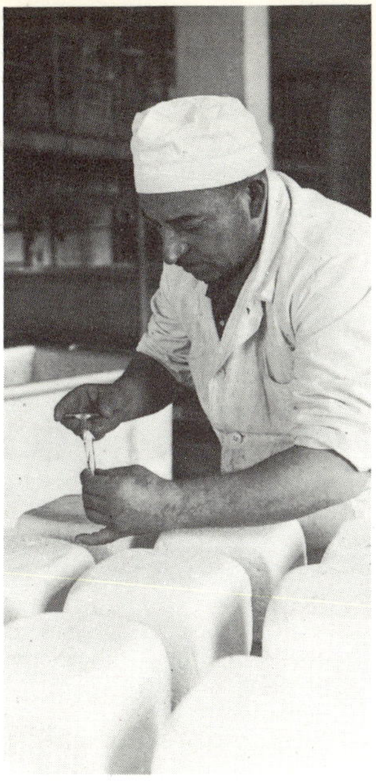

▲ Chanakh is Russia's best-known cheese; here a new batch is being "ironed" by one Khachik Akapyan, a Soviet cheese expert.

plastic-curd Caciocavallo or Provolone type. Interesting, but not by any means sensational, are the Burrinos—small, pale plastic-curd cheeses which contain a pat of sweet butter. A far more exciting discovery is Marscapone, the rich cream cheese often flavoured with crystallized fruit; or the rare Caprino; or genuine Pecorino Romano. And genuine Stravecchio—very aged Parmesan—is really worth searching out. The small piquant mountain Gorgonzolas and Mozzarellas made with buffalo milk are two great rarities now found only in some of the less frequented areas.

Something of the same is true in Greece; only in mountain villages are you likely to find real Fetta. Perhaps in the islands you may find something unique, but practically speaking, hygiene and indeterminate feedstuffs can make such experimentation too adventurous for any but the dedicated.

France, of course, offers the best pickings, especially if you are a goat's-cheese fanatic. Local cream cheeses too are fascinating. But remember that this is the twentieth century. What may be presented as a local speciality might be a recent invention—but may be none the less good for that.

As always, the inventive Americans offer much of interest. East Coast sage cheeses are delicious, and some American Cheddar types are exceptionally good. Many "copies" are actually authentic cheeses made by immigrant families or their descendants. Three indigenous cheeses are "brick" cheese, a small-holed, mellow variety; Liederkranz, a spreadable mild Limburger; and Monterey "Jack" cheese, a mild tangy white Californian cheese—a notable addition to Mexican dishes. South America offers many fresh white cheeses; those spiced with chilli are an experience. Argentina's large cheese production features quite an acceptable Parmesan-type.

A universal food

As well as watching out for new cheeses abroad, it is particularly interesting to look for different ways of using cheeses. Such ideas are easier and safer to transport home. Examples include Turkish and Greek squares of deep-fried pastry filled with cheese; Australian dishes with fruit and cheese; and South American fried hard cheeses spread with marmalade.

An interest in food makes any travelling that much more worth while. Local shopkeepers—language difficulties notwithstanding—will usually be delighted to tell you about their wares. Cheese is one thing almost everyone has in common. Use it as your passport to more interesting travel and more original travellers' tales.

Suppliers

It is always advisable to write for a catalogue and price list before ordering any equipment, as prices, availability and specifications may change.

GENERAL SUPPLIES

The first three can supply virtually everything you might need, including rennet. Those interested in real cheese starter should refer to the "Special supplies" section.

J. J. Blow Limited,
Oldfield Works,
Chatsworth Road,
Chesterfield, S40 2DJ
This company's most attractive product is the Dairythene $2\frac{1}{2}$-gallon polythene bucket with lid. Stainless steel bowls and buckets are available, too.

Clares Carlton Limited,
Wells,
Somerset, BA5 1SQ
The range includes the supplier's own press and Coulommiers molds, and also cheese record books. Very helpful people who can supply some interesting recipes, too.

Divertimenti,
68-70 Marylebone Lane,
London W1
Some of this company's equipment is specially made for it and it is expanding the range as demand increases. Ingredients sold include rennet and annatto. Unique in London.

Self-sufficiency and Small-holding Supplies Limited,
The Old Palace,
Priory Road,
Wells,
Somerset, BA5 1SY
The range includes an excellent Smallholder's Cheese and Wine Press, and a good range of molds. The company also publishes a self-sufficiency magazine. Recommended.

P.S. **Harrods Food Hall** sells rennet and annatto.

SPECIAL SUPPLIES

Please remember that proper cheese starter is expensive to buy and requires special equipment to incubate and store it, as well as a somewhat scientific mind. The following suppliers do not really cater for the small cheese-maker but you will always receive a courteous reply if you are practical in your enquiries.

Enolacto Limited
8 Hillfield Court,
London NW3
A wide range of cultures is available for all types of cheese, made by Laboratorium VISBY, Tønder, Denmark. There is a special phial of freeze-dried culture for the home cheese-maker, but even this costs well over £1. For the serious cheese-maker there is also a great deal of specialized equipment and test-kits for fat content, acidity, antibiotics, etc. Every culture comes with full instructions for its incubation.

R. J. Fullwood & Bland Limited,
Ellesmere,
Salop, SY12 9DG
This company specializes in cheese rennet and colouring in relatively small quantities, and can provide junket rennet too.

Chr. Hansen's Laboratory Limited,
476 Basingstoke Road,
Reading, RG2 0QL
This eminent company supplies freeze-dried cultures from Denmark for cheese and other dairy products, mainly for commercial manufacturers, but an enquiry may well prove worth while.

CALORIES AND YOUR DIET

Once you understand that fat content is a measure of the fat as a percentage of the dry matter of cheese, not of the cheese as a whole, your fear of cheese as a high-calorie food can be lessened.

It is impossible to give an accurate assessment of the calorie count of every cheese, because this depends on its condition, degree of dryness and so on. Generally, the wetter the cheese, the fewer the calories, for there is less solid matter containing fat. Therefore, a general rule is:
fresh soft cheeses: 100 calories per $3\frac{1}{2}$ oz/100 g
soft cheeses: 283 calories per $3\frac{1}{2}$ oz/100 g
hard cheeses: 368 calories per $3\frac{1}{2}$ oz/100 g.

As it is unusual to eat more than an ounce or two of cheese at a time you will see that few cheeses are likely to upset your diet very much. It is what you eat and drink *with* the cheese that should be watched more carefully.

Seven common cheese types:
(based on a $3\frac{1}{2}$-oz/100-g serving)

	FAT	PROTEIN	CALORIES
Blue cheeses	45%	21%	368
Camembert	50%	17-21%	286
Cheddar	48%	25%	398
Cottage cheese	4·2%	14%	106
Emmental	45%	28%	399
Gouda	48%	26%	365
Parmesan	40%	36%	390

Books

There are relatively few books on cheese and fewer on cheese-making.

When referring to recipes for cheese-making, remember that each recipe has been found to be successful in certain conditions, probably unspecified; this is why the recipes vary so much.

The best books for authentic cheese-making recipes are indubitably those you will find in second-hand shops.

The majority of the books listed here are either quite difficult or impossible to find outside public libraries, as they are out of print. Where available, current prices are given.

CHEESE APPRECIATION

Guide du Fromage, Aidan Ellis, England, distributed by Paxton & Whitfield and Foyles, £5.00.
For many people, this is the bible of cheese lore, but esoteric and inaccurate in the extreme and devoting only a small amount of space to English cheese.

The World Atlas of Cheese, Nancy Eekhof-Stork, Paddington Press Ltd, 1976, £12.95.
Big, expensive, somewhat inconsistent, but fascinating. A cheese book for the coffee table.

Cheese and Cheese Cookery, T. A. Layton, Wine and Food Society/Michael Joseph, 1967.
Excellent recipes and advice. A good, fairly basic but extensive guide to cheeses.

Choose Your Cheese, T. A. Layton, Gerald Duckworth and Co, 1957.
The seminal version of the above and somewhat superseded by advances in cheese technology.

The Cheese Handbook, T. A. Layton, Dover, 1975. £0.95.
A condensation of the 1967 book, updated. The real love of cheese oozes through every page.

The Cheeses and Wines of England and France, John Ehle, Harper and Row, 1972.
Written in typically rambling and turgid American style, but buried within is some really fascinating and usually practical information on cheese-making and wine-making. Worth having.

Cheeses of the World, André Simon, Faber and Faber, reprinted 1965.
An eminent author, but the age of the information shows somewhat. Good recipes, as may be expected.

Cheeses of the World, US Dept. of Agriculture, Dover Publications, 1972.
Detailed information on how many, many cheeses are made. Very little about how they taste.

The Cheese Handbook, Bruce H. Axler, Cassell, 1970.
Lots of information but little editorial consistency and confusing classifications. Interesting but self-indulgent section on cheese and wine, beer, etc.

A Little Book of Cheese, Osbert Burdett, Gerald Howe Ltd, 1935. Said to be the first book on cheese to be published in Britain.

Cheddar Gorge, *a book of English cheeses,* Collins, 1937.
A treasure to be snapped up if you are lucky enough to find one. A series of essays by famous people about each of the English cheeses. Informative and vastly amusing.

CHEESE-MAKING

The Backyard Dairy Book, Len Street and Andrew Singer, Prism Press.
Full of recipes, history and simple methods.

Home Cheesemaking, Susan Ogilvy. Batsford, 1976.
An interesting book but use proves it to be impractically planned. It relies mainly on laboratory cultures for starters and so requires a high degree of motivation and ability, in contrast to its apparently simple presentation.

The Story of Cheesemaking in Britain, Val Cheke, Routledge and Kegan Paul, 1959.
Very readable history since Roman times, with plenty of tips for home cheese-makers.

Food in England, D. Hartley, Macdonald, 1954.
A big totally absorbing book about food of every kind from way back. A few, but really wonderful, cheese recipes. A library may be your best bet, but buy it if you ever see it.

Dairywork for Goatkeepers, British Goat Society, 1973.
Lots of sound advice and practical recipes, including one for Stilton. Write to B.G.S., Rougham, Bury St. Edmunds, Suffolk 1P3 09LJ, enclosing 10p plus large s.a.e.

Soft Cheese, Ministry of Agriculture no. 458. Good, official advice, particularly if you want to sell fresh soft cheeses.

Cream Cheese, Ministry of Agriculture no. 222. Same usefulness as above.

Glossary

Acid: (descriptive of milk) synonym for sour, when milk lactose has been turned into lactic acid by the action of bacteria.

Acid-curd: curd formed by the natural action of bacteria or by adding an acid to fresh, sweet milk, e.g. lemon juice or vinegar.

Albumen: a water-soluble protein, a component of whey. It coagulates when heat is applied.

Ammoniacal: used to describe cheese, usually of the soft variety, which is overripe and smells of ammonia.

Annatto: cheese colouring, made from the pulp which surrounds the seeds of the West Indian tree of the same name.

Bleu: used two ways (a) to describe cheese with an internal mould (see *Persillé*) and (b) to describe goat's cheeses which have an external blue mould.

Bloom: the white, furry surface mould (*penicillium candidum*) such as that seen on Brie, Camembert and Neufchâtel.

Brine: a solution of salt and water in which cheeses are washed or soaked to encourage the formation of a skin or rind.

Brushed: a cheese which has been brushed during curing to control growth of surface bacteria (see also *washed*).

Casein: the major protein of milk which solidifies and precipitates when milk is coagulated by acid or rennet (see *curing*).

Cendré: means cindered; generic name for goat cheeses cured in ashes.

Cultured buttermilk: buttermilk is the by-product of churning cream to make butter. Buttermilk from raw cream will thicken naturally, because of bacterial action. Cultured buttermilk is made from pasteurized cream; a suitable culture is added (buttermilk starter) to restore these qualities. Cultured buttermilk is an ideal starter for cheese even though the bacterial family is not absolutely specific.

Curd: the solid matter or coagulum resulting from the formation of lactic acid in, or addition of rennet to, milk. This process is called curdling. The separation of the curds leaves liquid matter, the whey.

Curing: ageing or ripening of cheese. The action of different bacteria on treated curd results in the breakdown of casein. This produces a more digestible cheese, and causes side-effects such as formation of holes, aromas, characteristic surfaces and deeper flavour.

Double-cream (*double-crème*): a cheese containing 60-74% fat content.

Enriched: milk or milk-curd to which cream is added, to make double- or triple-cream cheeses.

Fermier: descriptive of a cheese made on a farm by traditional methods, usually from raw milk.

Flocculate: the formation of clouds of coagulated albumen as the result of applying heat to whey. This coagulated or flocculated protein is the basis for Ricotta and Mitzithra.

Full milk: milk as it comes from the cow, or any other animal, as far as cream content is concerned, i.e. milk from which no cream has been skimmed.

Homogenize: to break down the fat globules of full milk and distribute them evenly, so that the cream and milk do not separate. Such milk is an unsuitable product for cheese-making.

Lactic: descriptive of the smell and flavour of full milk, particularly when speaking of cheese.

Lactose: the milk-sugar. The action of bacteria on this substance creates lactic acid. Much remains in the curds and whey of renneted fresh milk, hence the sweetness.

Laitier: usually denotes a cheese made commercially, i.e. in a creamery.

Macerate: to steep in a liquid that one may also normally drink. With cheese this is usually something alcoholic, such as beer, cider or spirit. Note: brining or soaking in a salt solution is *not* macerating.

Mold/Mould: two meanings, two spellings. In this book,

"mold" is used for the container in which cheese is formed or shaped: "mould" means the bacteria which grow on the surface of some cheeses (see *bloom*) or in the body of other cheeses—"blue" mould. The external mould is usually *penicillium candidum*, which produces white furry cheeses. The most common internal mould is *penicillium glaucum* or *penicillium roqueforti*.

Pasteurization: the partial sterilization of milk by heating it to 160°F (71°C) for 15 seconds and cooling rapidly. Pasteurization kills all disease-carrying organisms together with most of the bacteria which would help in natural curd formation.

Persillé: usually used to denote a blue-mould cheese made from sheep's or goat's milk, e.g. Roquefort. It derives from the French word for parsley and refers to the appearance of the mould.

Rancid: descriptive of the rank, unpleasant, oily taste arising from the oxidation of butterfat in cheese or butter. Note: rancidity spreads like wildfire; get rid of anything rancid as soon as you discover it or it will affect perfectly fresh butter and cheese.

Raw: descriptive of milk which has not been subjected to any treatment for the removal or control of bacteria and so contains its full quota of naturally-occurring bacteria. Many scientists feel that no raw milk is absolutely safe straight from the cow. It is sometimes advised to wait until it is beginning to sour before drinking it—the lactic acid will have killed many harmful germs.

Rennet: a naturally-occurring substance that curdles milk without souring it, although the presence of some degree of sourness or acidity helps its action. Usually made from the fourth stomach of young ruminating mammals, such as calves or kids.

Rennet substitutes: there is no artificial rennet, but the juice of several flowers and plants works in a similar way.

Skimmed milk: milk from which all or some of the cream has been removed. It is obtained either by standing the full milk overnight and ladling off the cream, or by putting it into a centrifugal separator which works instantly. The second is the most usual method today.

Starter: a controlled growth, in previously sterilized milk, of those bacteria or groups of bacteria which are required to replace those destroyed by pasteurization. Modern science has developed starters specifically intended for different kinds of cheeses, as well as for butter, yoghurt and buttermilk. The growth and use of cheese starters is complicated and precise—and can be expensive. Very good results are obtained by using cultured buttermilk. In recipes in this book which use buttermilk, starter can be substituted, in a similar quantity.

Sterilized milk: milk treated at over boiling point for up to half an hour. It is usually homogenized first. Virtually all micro-organisms are destroyed and the treatment does alter the colour and flavour of the milk. Unsuitable for cheese-making.

Triple cream (*triple-crème*): cheese with a fat content of 75% or more.

Truckle: a term used for smaller versions of pressed cheeses. A farmhouse Cheddar may weigh 60 lb (24 kg). A truckle Cheddar may weigh as little as 2 lb (·8 kg) or as much as 15 lb (30 kg).

UHT: ultra-heat-treated milk, raised to a very high temperature for a very short time then sealed in aseptic containers. Keeps without refrigeration for months. Is said to taste the same as pasteurized milk. Unsuitable for cheese-making.

Washed: the rinsing of the rinds of soft cheeses with water or brine to inhibit the growth of *penicillium candidum* mould, but encourage the growth of bacterial cultures—usually red coloured.

Whey: the green-tinted, clear liquid formed when milk coagulates. It may be slightly clouded with butterfat when first you cut curd in your cheese-making. If the cloudiness continues, you have cut the curd too soon or are handling it too roughly. This results in a less tasty and much tougher cheese.

Index

Numbers in italics indicate illustrations.

Acid-curd cheeses, 26, 40, 64, 80, 92
Acidometer, 38
Albumen cheese, 28, 45, 64, 92
Alcohol, dosing cheese with, 52
All-American cheesecake, 60–61
Annatto, 35, 36, 92
Appenzell, 84, 85
Apple and cheese pie, 62–63
Argentinian cheese, 22
Aurochs, 8, 12
Ayrshire cows, *9*

Backyard cheese, 44
Baked apple with grated cheese, 62
"Banbury cheeses", *15*
Banon, 45, 65, 70
Beaufort, 70, 88
Bel Paese, *25*, 82; Crema, 82
Belgian cheese, *23*, 88
Berkeley, 79
Beverages with cheese, 53–55, *53*, *55*
Black cherry cheesecake, *56*
Bleu d'Auvergne, 70
Bleu de Bresse, *25*, 70
Bleu de Causses, 70
Blue cheeses, *4*, 6, *25*, *27*, 28–29, 48, 55, 64, *69*, 74, 81, 86, 87, 90
Blue Cheshire, *4*, *25*, 54, 55, 74, 76, *77*, 79
Blue Stilton, *25*, *53*, 79
Blue Vinny, 76
Blue Wensleydale, 76, 79
"Bluing", *11*, *27*, 28–29, 33
Bonbel, 70
Bondon, 70
Boursault, 70
Boursin, 24, *25*, *69*, 70
Brahmin cattle, 7, *7*
Braunvieh cows, *9*
"Brick" cheese, 89
Brie, *5*, 13, 20, 26, 51, *52*, *53*, 54, 55, 65, 66, 68, *69*, 70–71, 78; *Fermier*, *11*, 32, 71; *Fermier de Meaux*, *25*, *66*, 71; Irish, 22; *Laitier*, *25*, 70–71
Brie tart, 57
British cheeses, 74–79
British White cows, 8
Brown whey cheeses, *25*
Buffalo cheeses, 7, 89
Burrino, *29*, 89
Buying cheese, 46–47, *49*

Caboc, 79
Caciocavallo, 7, 27, 81, 89
Caciotta, 27, 81
Caerphilly, 26, 44, 55, 78; Farmhouse, *53*, 54, 76, 78
Calories, 10, 90
Cambridge or York cheese, 43, 74, 78
Camel cheese, 8
Camembert, 17, *25*, 26, 51, *52*, *53*, 55, 57, 64, 65, 66, 67, *67*, *69*, 78, 90; *Fermier*, 64, 71; *Laitier*, 71
Canadian cheeses, *17*, 22, *23*
Cantal, *69*, 70, 71
Caprice des Dieux, 71
Caprino, 81, 89
Carré de l'Est, 71
Castello, 87
Celery stuffed with Stilton, 63
Cendré goat's cheese, *25*, 71, 92
Chabichou goat's cheese, 68, 71
Chanakh, *89*
Cheddar, 10, *11*, 20, 27, *52*, 55, 62, 65, 66, 74, *77*, 78, 90; American, 89; Canadian, 22; English Farmhouse, 23, 74; Farmhouse, 32, *33*, *46*, 47, *53*, 67, *77*; French, 22, 74; giant, *19*; New Zealand, 74; Scottish Farmhouse, *5*
Cheese biscuits, 62
Cheesecakes, 59–61; all-American, 60–61; black Cherry, *56*; fragrant, 59; Käsekuchen, 60; miniature, 59–60; with yoghurt, 60
Cheese cubes, in salad, 62
Cheese loaf, *56*, 58
Cheese presses, *18*, *32*, *74*; improvised, *42*
Cheese rarebits, 62, 63
Cheese scones, *56*, 62
Cheese soufflé, 63
Cheese-stuffed potato, *56*, 62
Cheese-tasting, *11*
Cheesecloths, 27, *31*, 36, *39*
Cheshire, *11*, 32, 74, *75*; Blue, *4*, *25*, 54, 55, 74, 76, *77*, 79; Red, *25*, *77*, 78; White, 26, 62, 79
Chester, *75*
Chèvres, *9*, 13, *25*, 45, *53*, 54, *69*, 71
Chèvreton d'Ambert, 71
Chevrotin, 71
Coagulation, 38, 40
Coeur à la crème, *56*
Colander, 36, *39*
Colwick cheese, 42–43
Cottage cheese, 10, 24, 26, 90
Cottenham, 76
Coulommiers, 71–72; on straw, *25*, 71
Coulommiers hoops/molds, 36, *39*, 42
Coulommiers-mold cheese, 42
Cows, 8, *8–9*, 14
Cream cheeses, 24, *25*, 43, 50, 62, 63, 64, 74–75, 89; double-, 41–42, 92; single-, 41
Crema Bel Paese, 82
Crottin, *25*
Curd cheeses, 24, 40–41; acid-, 26, 40; lactic, 40; rennet-, 26, 41
Curd-cutting, 38
Curds and whey, 6, *6*, *14*, 92, 93
Curing cheese, 18, 26, 92
Cutting cheese, *50*, 51

Dairy Shorthorn cows, *9*
Danablu (Danish Blue), *27*, 86, 87
Danbo, 87
Danish cheeses, 22, *23*, *27*, 80, 86, 87, *87*
Derby, 77, 79; Sage, *25*, 76, *77*, 79
Dolcelatte, 47, 82
Double Gloucester, 8, *25*, 76, 79
Drainage boards and mats, 36, *39*
Drying room, *18*
Dunlop, 77–78, 79
Dutch cheeses, 23, *23*, 33, 83, *83*, 84, 85, 88

Edam, 10, *25*, *83*, 84, 85
Egg cheese, 57
Elbo, 87
Emmental, *25*, 26, 27, *30–31*, 62, *65*, *69*, 70, 78, 84, 85, 88, 90

Equipment (for home cheese-making), 36, *37*, *39*
Esrom, 87
Exports of cheese, 16, *20–21*, 21, 22, 23, *30*, 70, 83, 84

Farmhouse cheeses, *5*, *25*, 32–3, *33*, *53*, 67, 76
Fat content, 64, 90
Fermiers, 32–33, 64, 92
Fetta, 23, 29, 32, 58, 62, 66, 80, 89
Finnish cheese-making, 23, *23*
Fiore Sardo (Pecorino Sardo), 82
Fish pie, *56*, 63
Flavour variations for soft cheeses, 41, 61
Fondu cheeses, *25*, 29, 73
Fondue, *56*, 63
Fontainebleau, 54, 72
Fontina (Fontal), 82
Fougeru, 72
Fourme d'Ambert, *25*, 72
French cheeses, 7, *9*, 22, 23, *23*, 32–33, 43, 66–73, 74, 88, 89
Fresh cheeses, 24, *25*, 26, 33, 88, *88*, 89
Friesian cows, 8, *8*, 83
Fromage blanc, 72
Fromage de Monsieur, 72
Fromage en pot, 61
Fromage fort, 61; *à la lyonnaise*, 61

Galotiri, 28
Gammelost, 88
Gaperon d'Auvergne, 54, 72
Garlic and herb cheese, *25*
German cheeses, *16*, *18*, 21, *23*, 83, 84, 85, 88
Gjetost, *25*, *86*, 87
Gloucester cows, 8
Goats, goat-milk cheeses, 7, *9*, 14, *25*, 29, 33, 35, 44–45, 48, 54, 68–69, 81, *88*, 89
Gorgonzola, *25*, 81, *81*, 82, 89
Gouda, 10, *11*, 13, 27, 65, 66, 78, *83*, 84, 85, 90; Farmhouse, 32, 33; Mature, *75*
Gouda cornets, *56*, 63
Gougère, *56*, 58; cake, 58; hors d'oeuvre, 58
Grana, 81, *81*, 82
Greek cheeses, 7, *16*, *17*, 23, *23*, 28, 29, 32, 62, 80, 89
Greek salad, *56*
Gruyère, *25*, 26, 27, 48, 54, 58, 62, 63, *65*, *69*, 70, 78, 84, 85, 88

Handkäses, 21, 84, 88
Harel, Mme Marie, *67*
Havarti, 87

Ilchester, 79
Indian cheese, 22
Irish cheeses, 22, 74, 78
Italian cheeses, 7, *9*, *16*, 23, *23*, 29, *29*, 62, 80–82, *81*, 88–89

"Jack" cheese, Monterey, 89
Japanese cheeses, 22
Jarlsberg, 86, 87
Jersey cows, 8, *9*, 22, 35

Käsekuchen, 60
Kerry cows, 8
Kingston cheese, 43–44

Lactic acid, 5, 26, 35
Lactic curd cheese, 40
Lactic goat-milk and sheep-milk cheeses, 44–45
Lancashire, 62, 79
Larzac sheep, *9*
Leicester Red, *25*, *53*, 79
Leiden cheese, 85
Liederkranz, 89
Limburger, 84, 85, 88, 89
Liptauer, 61–62
Livarot, *25*, 26, 72
Llama cheese, 8
Low-fat cheeses, 10

Macaroni cheese, *56*
Macerated cheeses, 68, 70, 92
Mainzerkäse, 85
Maroilles, 26, *69*, 72
Marscapone, 89
Maturing of cheeses, *11*, 29, 38
Measuring jugs, 36, *39*
Mi-chèvre, *25*, 70, 72
Milk, acidity, 38; for home cheese-making, 35; types, 64
Mitzithra, 28, 32, 80
Molds, 36, *39*, 63, 92–93
Mongolian cheeses, 22
Mozzarella, 22, *25*, 82, 89
Mozzarella and orange salad, *56*
Münster (or Munster), 26, 72, 84, 85
Mycella, 87
Mysost, 87

Nagelkaas, 88
"Needling", *27*, 28–29
Neufchâtel, *25*, 43, 69, 72
New Zealand cheese, 17, 21, 22, 23, 74
Normandy cows, *8*
Norwegian cheeses, 18, 21, 23, 86, *86*, 87, 88

Orkney and Islay cheeses, *34*, 78, 79

Parmesan (*Parmigiano*), *11*, *13*, 22, 48, 62, 81, *81*, 82; *Giovane*, 82; *Tipico*, 82; *Stravecchio*, 82, 89; *Vecchio*, 82
Pasta filata, 81
Pasteurization, 5–6, 18, 26, 30, 34, 64, 65, 67, 93
Pavé, 73
Paxton and Whitfield, *46*
Pecorino, 7, *9*, 29, 62, 81, 82; Romano, 82, 89; Sardo, 82
Pennard Giant cheese, *19*
Petit Pont l'Evêque, *25*
Petit Suisse, 65, *69*, 72–73
Philadelphia cream cheese, 60
Pineapple cheese, 22
Pipo Crèm, *52*
Pizza, *56*
Plastic-curd cheeses, *25*, 27, *29*, 64, 81, 89
Poivre d'Auvergne, *25*, 73
Pont l'Evêque, 26, *52*, *53*, 73
Port (du) Salut, 15, *25*, *69*, 70, 73
Potte Kees, 61
Potted cheese, *56*, 61–62
Pre-packed cheese, 47
Pressed cheeses, 17, *17*, 22, 24, 26–27, *27*, 28, 29, 38, 40, 43–44, 48, 55, 74
Processed cheese, 13, 22, *25*, 29, 64
Provolone, *13*, 27, 81, *81*, 82, 89

Quark cheese, 84, 88
Quark tart, *56*, 59

Raclette, *51*, 88
Reblochon, *25*, 26, 73
Reblochon Fermier, 23

Red Cheshire, *25*, *77*, 78
Red Leicester, *25*, *53*
Red Windsor, 79
Reindeer cheese, 7
Rennet, 5, 6, 12–13, 14, 18, 26, 35, 36, 38, 40, 75, 93
Rennet-curd cheeses, 26, 41, 64
Ricotta, 28, 45, 80, 82
Ripened cheeses, 26–29
Ripening cheese, 6–7, 40, 52, 64–65
Roman cheese-making, 14–15, 47, *75*, 80–81
Roquefort (cheese), *4*, 7, *9*, 13, *25*, 29, 33, *53*, 54, 57, 66, *68*, 70, 73
Roquefort sheep, *28*
Russian cheeses, 21, *23*, *89*

Sage cheeses of US East Coast, 89
Sage Derby, *25*, 76, *77*
Sage Lancashire, *25*, 76
St Marcellin, 73
St Nectaire, 69, 73
St Paulin, *69*, 73
Sainte-Maure, 71
Salting, 6, 35
Samso, 87
Sauerkraut toast, 62
Scalded cheeses, 26–27, 28, 64
Scandinavian cheeses, 86–87, 88
Schabzieger, 28, 88
Scottish cheeses, *5*, 77–78
Sheep, sheep-milk cheeses, 7, *9*, *28*, 29, 35, 44–45, *88*; lactic, 44–45
Single Gloucester, 79
Soft cheeses, 22, 40–43, 48, 74–75, 77
Soft paste cheeses, 26, 64
Sopravissina cows, *9*
Soup with grated cheese, 63
South American cheeses, 22, 89
Spiced cheeses, 61–62
Spinach and cheese pie, 58–59
Spinach quiche, *56*
Starter cultures, 6, 18, 26, 35, 93
Sterilization, 18, 38, 40, 93
Stilton, *4*, *11*, 52, *52*, 54, 55, 57, 65, 68, 74; Baby, *11*; Blue, *25*, *53*, 79; New Zealand, 22; White, 58, 62, 79
Stilton-stuffed celery, 63
Storing, 48
Stuffed vegetables, 63
Suffolk cheeses, *15*
Swedish cheeses, *23*, 86
Swiss cheeses, *23*, *25*, 28, *28*, *30–31*, *51*, 83–84, *83*, 85, 88

Taleggio, 54, 82
Tests and techniques (in home cheese-making), 38–40
Tomme(s), 29, 54, 73; au Raisin, 29, *52*
Tour Grise, 73
Turkish cheese, 27, *88*
Tybo, 87

United States cheeses, *17*, *19*, 21, 22, *23*, 89, *89*

Vegetable rennet cheese, 47
Vegetables with grated cheese, 63
Vegetarian cheese, 47

Walnut cheese, 29
Walton, 79
Washed rind cheeses, 68
Welsh Black cows, 8
Wensleydale, Blue, 76, 79; White, 79
Whey cheeses, 21, 27-28, 45, 64, 80
Whey collector, 36
White Cheshire, 26, 62, 79
White Stilton, 58, 62, 79
White Wensleydale, 79
Windsor, Red, 79
Wine and cheese, 53-55, *53*, *55*, 67, 70, 76-77, *81*

Yoghurt, cheesecake made with, 60
York cheese, *see* Cambridge

Zebu (Brahmin cow), 7

Credits

Artists
Ron Hayward Art Group
Vanessa Luff
QED
John Shackell

Photographs
Heather Angel, 4
Bavaria Verlag, 3 (top left), 30 (top right, bottom left, bottom right), 31 (bottom left, bottom right)
Danish Centre, 27, 87
English Country Cheese Council, 75 (bottom right), 78
Robert Estall, 2 (bottom right), 3 (bottom left), 5, 11 (bottom left), 19, 28 (right), 29, 32, 33, 37, 65, 75 (top right), 77
Mary Evans Picture Library, 66
Paul Forrester, 2 (centre right), 25, 40, 41, 42, 48, 49 (top), 52, 53, 56
French National Tourist Office, 28 (top left), 68
Sonia Halliday, 88
Robert Harding, 12, 45
Hickory Farms, Inc., 89 (bottom)
Mansell Collection, 15
Dennis Moore, 2 (top right), 83 (bottom left)
Margaret Murray, 7
Norwegian Dairies Association, 86
Novosti, 89 (top left)
Paxton and Whitfield, 46
Radio Times Hulton Picture Library, 14, 18 (bottom)
Swiss National Tourist Office, 11 (bottom right), 30 (centre), 31 (top left, top right, centre right), 83 (top right)
SEF, 13
Suddeutscher Verlag, 84
John Topham Library, 11 (top right), 18 (top left), 34, 49 (bottom), 51, 67, 69, 74

Materials for photograph on page 37 kindly loaned by Self-sufficiency and Smallholding Supplies Limited, Wells, Somerset.

Cover
Design: Design Machine
Photograph: Paul Forrester